The Greek Seashore
A Field Guide to
Coastal Invertebrates

I. E. BATJAKAS - A. E. ECONOMAKIS

I. E. BATJAKAS & A. E. ECONOMAKIS

The Greek Seashore

A Field Guide to
Coastal Invertebrates

Photographs by
Alistair E. Economakis

Drawings by
Ioannis E. Batjakas

EFSTATHIADIS GROUP

EFSTATHIADIS GROUP S.A.
14, Valtetsiou Str.
106 80 Athens
Tel: (01) 5154650, 6450113
Fax: (01) 5154657
GREECE

ISBN 960 226 601 5

Printed and bound in Greece

Acknowledgements

The authors would like to thank all the people who assisted in the preparation of this book. The Psariotis family from Linaria, Skyros provided some specimens. We are grateful to members of our families, for their help and moral support. These include Alex, Tara, Adrian and Jessie Economakis, Jessie Phillips, Eystratios and Eystathia Batjakas, as well as Angela Batjakas Vatou. We would also like to thank the following people for their valuable help in collecting specimens, proof reading and assisting in the identification of some of the specimens: Angie Athanasiadis, Eric Derfner, Angeliki Evgenidou, Julie Fricke, Amy John, Les Kaufman, Jae Kim, Jasmine Maihta, Maria Mallidis, Rachael Max, Alex McGregor, Alexandra Mela, Dan Painter, Giorgos Panayiotou and Giorgos Skaltsis.

Contents

Introduction

Since prehistoric times, marine invertebrates have been a vital resource for many coastal settlements due to the fact that several groups of marine invertebrates inhabit shallow environments and are easy to collect. For these reasons, marine invertebrates have played important roles in societies which have looked to invertebrates as a relatively easily collectable source of food (protein), decoration, and currency for thousands of years.

Greek civilization and culture has always been strongly associated with the sea. The sea provided food, a source of income, communication and trade routes. Together with fish, marine invertebrates often appeared on ancient Greek art. Greeks still value marine invertebrates as a source of food and income. For example, many of the edible invertebrates (e.g. lobsters, crabs, octopus, shrimps) attain high prices at the Greek seafood markets, and on Kalimnos Island in the Aegean Sea, the economy is based on sponge collecting.

Marine invertebrates are still used today as a food resource and are valued in the jewelry industry (everyone can appreciate the value of pearls!) and most recently are becoming important in the biochemical industry; it has been determined that certain sponge species have been found to posses important antibiotic and anti inflammatory compounds which are used in some modern medicines.

This guide is intended to compliment our previous book, "Coastal Fishes of Greece". Similar to "Coastal Fishes of Greece" in its precise, scientific accuracy, "The Greek Seashore" focuses on the study of many common invertebrates found in coastal Greek waters. Included in the guide is general information about the relative phyla and classes, the scientific names (Genus species), classification information, life histories and phylogenetic relationships of each group of organisms as well as the description of each species which includes the common names in English (En), Greek (Gr) and German (G)[1]. These are followed by classification information and general comments on palatability or potential health hazards[2]. More than 80 species and genera representing 11 phyla are described providing tourists, divers, sport fishermen and naturalists alike with one of the most complete guides available.

---

[1]Readers should note that common names can vary depending upon location.
[2]Although we tried to use the most accepted classifications for the described species, it should be noted that classifications for many invertebrates have not yet been standardized within the scientific community.

Of the over 1 million described animal species, only approximately 50,000 are vertebrates. The other 950,000 species are invertebrates belonging to nearly 40 phyla. There is huge variation in morphology and life history patterns not only across, but also within invertebrate phyla which makes the group of organisms a fascinating study.

The authors referred to the following literature. All are useful to anyone who is interested in learning more about marine invertebrates. Some of the drawings are modified and redrawn from Ruppert & Barnes, 1994.

Colin, P. L. 1978. Caribbean Reef Invertebrates and Plants. T. F. H. Publications Inc. Ltd. Neptune City, NJ, U.S.A. pp 512.

D'Angelo, G. & S. Gargiullo. 1979. Guida Alle Conchiglie Mediterranee: Conoscerle Cercarle Collezionarle. Fabri Editori. Milano, Italy. pp 223.

Gosner, K. L. 1978. Atlantic Seashore. Peterson Field Guides ®. Houghton Mifflin Co. Boston, U.S.A. pp 329.

Hayward, P. J. & J. S. Ryland (eds). 1995. Handbook of the Marine Fauna of North-West Europe. Oxford University Press. Oxford, England. pp 800.

Imai, T. (ed.) 1980. Aquaculture in Shallow Seas: Progress in Shallow Sea Culture. A. A. Balkema. Rotterdam, Netherlands. pp 615.

Martinez, A. J. & R. A. Harlow. 1994. Marine Life of the North Atlantic. Norman Katz. MA, U.S.A. pp 272.

Moosleitner, H. & R. Patzner. 1995. Unterwasserführer Mittelmeer Niedere Tiere. "Underwater Guide Mediterranean Invertebrates". Delius Klasing. Edition Naglschmid. Stuttgart, Germany. pp 214.

Morse, E. D., Chew, K. K. & R. Mann (eds.). 1984. Recent Innovations in Cultivation of Pacific Molluscs. Proceedings of an International Symposium Sponsored by the California Sea Grant College Programs in Alaska, Hawaii, Oregon, and Washington. Elsevier. Amsterdam, Holland. pp 404.

Muir, J. F. & R. J. Roberts (eds.). 1982. Recent Advances in Aquaculture. Croom Helm. London, England. pp 453.

Nordsieck, F. 1968. Die europäischen Meeresmuscheln (Prosobranchia). Vom Eismeer bis Kapverden und Mittelmeer. Gustav Fisher Verlag. Stuttgart, Germany. pp 273.

Nordsieck, F. 1969. Die europäischen Meeresmuscheln (Bivalvia). Vom Eismeer bis Kapverden, Mittelmeer und Schwarzes Meer. Gustav Fisher Verlag. Stuttgart, Germany. pp 256.

Nordsieck, F. 1972. Die europäischen Meeresmuscheln (Opisthobranchia mit Pyramidellidae; Rissoacea). Vom Eismeer bis Kapverden, Mittelmeer und Schwarzes Meer. Gustav Fisher Verlag. Stuttgart, Germany. pp 327.

Parenza, P. 1970. Carta d'identit_ delle conchiglie del Mediterraneo. Vol 1 Gasteropodi. Ed. Bios Taras. Taranto, Italy. pp 283.

Pechenik, J. A. 1996. Biology of the Invertebrates. Third Edition. Wm. C. Brown Publishers. Dubuque. IA, USA. pp 554.

Pietra, F. 1990. A Secret World. Natural Products of Marine Life. Birkhauser Verlag. Basel, Switzerland. pp 279.

Ruppert, E. E. & R. D. Barnes. 1994. Invertebrate Zoology. Saunders College Publishing. Fort Worth, TX, U.S.A. pp 1056.

Tenekidis, N. S. 1989. Mia Syllogi Koghylion apo tis Ellinikes Thalasses. "A Collection of Shells from the Greek Seas". Protopapa Brothers Ltd., Greece. pp 187.

Wood, E. M. 1983. Reef Corals of the World. Biology and Field Guide. T.F.H. Publications, Inc., Ltd. NJ, U.S.A. pp 256.

Marine Environments

Marine invertebrates are very diverse and can be found in a great variety of aquatic environments. The three main benthic habitats in coastal waters of Greece are: rocky, eelgrass *(Posidonia)* and sandy. Many invertebrates have a strong association with their environment. This relationship is dictated by the behavioural and ecological requirements of each species, and in many cases is so strong (especially for the sessile invertebrates) that it is unusual not to find some species in a particular habitat.

Some of the most common invertebrates that one can find in rocky environments are limpets, abalones, oysters, periwinkles, crabs, sea urchins, fire worms and sea anemones. Even though soft bottom habitats (sandy or muddy) do not provide as many microhabitats as other environments, they are ideal for (partial or total) burrowing organisms and most of their predators. The fauna in these environments is dominated by burrowing bivalves (eg. razor clams, cockles, clams), annelids (eg. feather duster worms, acorn worms) and echinoderms (eg. sand dollars, some star fish, sea cucumbers). The major groups of invertebrates that can be found in eelgrass environments are crustaceans (eg. shrimp, amphipods, isopods), bryozoans (these usually encrust the blades of eelgrass), some sea urchin species and some annelids.

Other organisms are found on the transition zone from one environment to another. For example, the pen shell is usually found in the sparsely vegetated transition zone between sandy and seagrass environments. Certain invertebrates, especially mobile predators/scavengers such as some cephalopods and crustaceans, can be found in all the different types of environment.

Rocky Environments

Eelgrass Beds

Sandy Environment

Underwater Cave in Rocky Environment

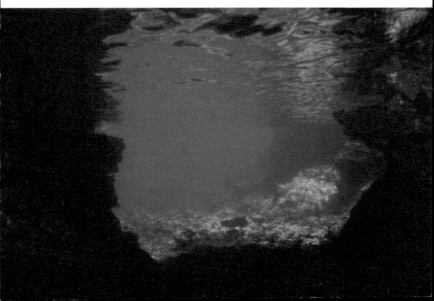

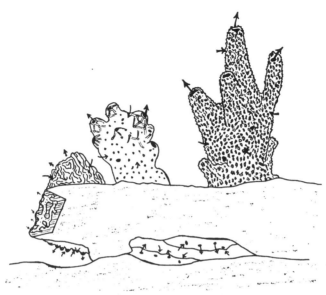

Different forms of sponges.
Arrows indicate water movements in and out of the sponges.

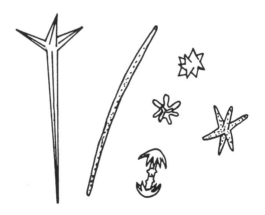

Different sizes and shapes of sponge spicules.

Phylum Porifera (sponges)

Sponges are the most primitive single celled organisms. Most species are marine and can be found in all seas on most types of substrates. Even though the majority of sponges are found in shallow waters, some species live at great depths. Sponges are sessile and have a porous body that lack organs. They respire and feed by passing water through a system of internal canals using flagellated cells. As the water passes through the canals, food particles are retained and digested by specialized cells. Even though most sponges obtain nutrients by filtering tiny particles from the water, some sponges obtain a percentage of their nutrients from symbiotic cyanobacteria; while others trap small crustaceans on their surface and slowly digest them. Sponges can filter large quantities of water; for example, a sponge 10 cm high can pump 22 liters of water through its body in 24 hours. Usually the smaller and more numerous pores are incurrent pores and the larger, less numerous, are the excurrent pores. The skeletal structure is complex and usually made up of a variety of sizes and shapes of spicules. Spicules, which are used for species identifications, can be made up of calcium carbonate, silica, or spongin. Sponge skeletons are composed of one type of spicule or a combination of siliceous and spongin spicules.

Sponges reproduce sexually and asexually; although most species are hermaphroditic, some are dioecious. During sexual re-production, males release sperm into the water column. The sperm eventually enters an incurrent channel of a female. Once inside the female, the sperm are transported to the eggs. In most species, the eggs develop inside the sponge; however, in some species belonging to the class Demospongiae, the fertilized eggs are released into the water column. In at least one species fertilization is known to be

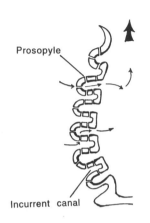

Prosopyle

Incurrent canal

Syconold type sponge

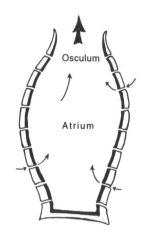

Osculum

Atrium

Asconoid type sponge

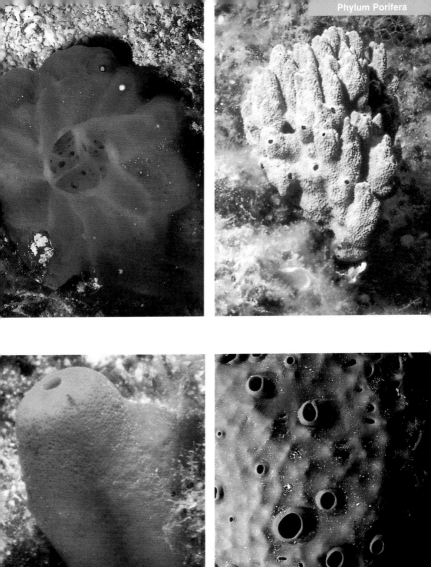

external. The eggs develop into larvae which usually have a brief free swimming stage before settlement. During asexual reproduction, sponges may form and release either groups of essential cells or (more rarely) buds.

Sponges provide a home for a variety of organisms, mainly invertebrates. Many sponges protect themselves from potential predators and compete with other organisms by producing metabolites. Secondary compounds from some species of sponges have been found to have antibiotic and anti-inflammatory properties. For these reasons, sponges are being investigated by biochemical industries for their medicinal uses. Sponge fishing for bathing sponges plays a dominant economic role for some Greek islands (e.g. Kalimnos). Commercial sponges have been overexploited in Greek waters and because of this, the majority of sponge fishing now occurs in the southern Mediterranean.

The phylum Porifera is composed of four classes:

Class Calscispongiae or Calcarea (calcareous sponges)

The spicules of these sponges are made up of calcium carbonate and are usually similar in size and separate. Sponges from this class can occur in a variety of colours such as red, yellow and purple; however many species are dull coloured. Most inhabit coastal shallow waters and reach a height of less than 10 cm.

Class Hyalospongiae or Hexactinellida (glass sponges)

These sponges are generally found at depths greater than 200 m. They are pale in colour and usually 10 to 30 cm tall. This class contains the most symmetrical sponges (usually they are vase shaped). Some form very beautiful intricate structures. The spicules often fuse together forming long strands made of silica. These

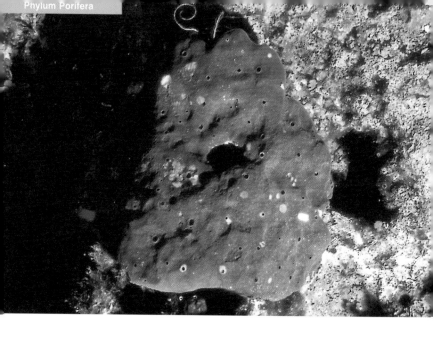

strands resemble opaque glass fibers, thus the name glass sponges. Some sponges from this class *(Euplectella spp.)* have an interesting symbiotic relationship with some shrimp species of the genus *Spongicola*. A male and female shrimp enter the sponge at a young age and as they grow they eventually become too big to leave the sponge. The shrimp pair live the rest of their lives trapped inside the sponge, feeding on plankton. In this way they are protected from some predators and are assured faithful mates.

Class Demospongiae (boring sponges, bath sponges)

This is the largest class of sponges containing approximately 90 % of all sponge species. Having the widest range, this class can be found from intertidal to abyssal depths worldwide. This class also

contains the only families of freshwater sponges. Usually demo-sponges are brightly coloured. Members of this class are very variable in shape and their size may vary from less than 1 mm to more than 2 m in diameter. The spicules from these sponges are mainly made up of silica or spongin. Some genera have spicules made up of both silica and spongin and one genus *(Oscarella)* does not have a skeleton at all. This class contains the boring sponges, which are able to bore through calcareous structures such as corals and shells. Bath sponges containing spongin fibers are also from this class. These sponges are harvested by Greek divers and on some islands, such as Kalimnos, sponge collecting plays a central role in the economy and the culture of the islanders.

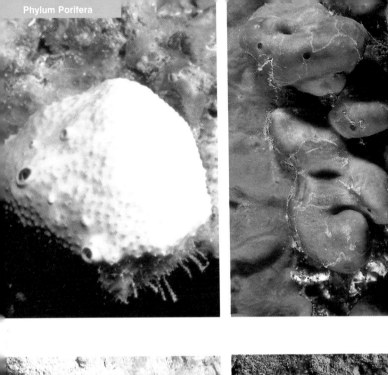

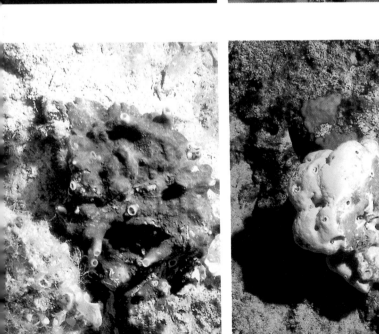

Class Sclerospongiae (sclerosponges)

This is a small class of sponges that usually inhabit caves and deep crevices in coral reefs. They have spicules composed of silica and spongin. These sponges have a calcium carbonate skeleton, and are the only organisms that deposit both silica and calcium carbonate. Sclerosponges can grow up to 1 m in diameter.

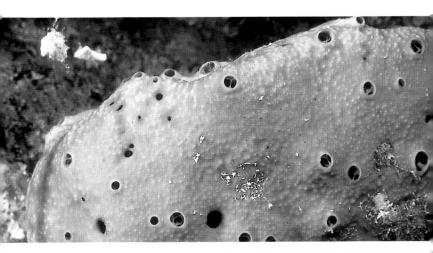

Medusoid body form

Polypoid body form

Phylum Cnidaria (jelly fish, corals and hydroids)

This aquatic phylum consists of approximately 9,000 species, most of which are marine. Cnidaria are radially symmetrical organisms with a single gastrovascular cavity, known as the coelenteron. Food enters this cavity from the mouth and is digested and absorbed. There is no anus. Organisms in this phylum are generally characterized by one of two body forms: the polyp and the medusa. Polyps are usually cylindrical and sedentary; most are colonial. One end of the polyp is generally attached to the substrate and the opposite end possesses a mouth surrounded by tentacles. The medusa form is usually free swimming and possess a body ranging in shape from a bell to a saucer. The mouth is located on the concave end and is surrounded by tentacles. Many cnidarians can alternate body forms from one generation to another. Most cnidarians are carnivorous and feed mainly on zooplankton. Some obtain a portion of their nutrient requirement with the help of symbiotic algae (zooxanthellae). Nematocysts are characteristic of cnidarians. These are microscopic more or less oval shaped capsules that contain a long coiled tube with a barbed end. When a nematocyst is triggered, the barbed end is ejected from the capsule and penetrates the surface tissue of the prey. Toxin is then ejected through the attached tube into the prey's tissue. Nematocysts are not only used for capturing prey, but are also used for defense. Cnidarians can reproduce sexually or asexually depending on their body form. In general, polyps reproduce asexually and medusas reproduce sexually.

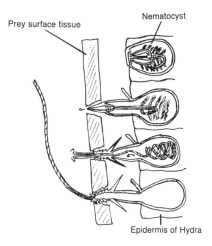

Discharge of a cnidarian nematocyst into prey

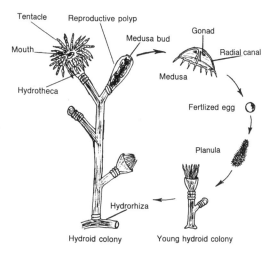

An example of a hydroid life cycle

Class Hydrozoa (hydroids)

There are approximately 2,700 species of hydrozoans, including all the freshwater cnidarians. Hydrozoans usually posses both a polyp and medusa form. Some can posses both during a life cycle, whereas others may only be able to display one form. Although the medusa form is easily identifiable, the polyp colonies can easily be mistaken for algae by the casual observer. Some species have a calcareous skeleton and can form structures similar to those of corals. Hydroids, with the exception of some species, such as hydras, are colonial. These colonies are made up of polyps attached to each other, with a continuous gastrovascular cavity. They are usually sessile, attached to the bottom by a rootlike structure called a hydrorhiza. Some hydroid colonies such as the Portuguese-man-of-war are pelagic and consist of both polyploid and medusoid forms. Hydroid colonies are usually polymorphic and each member has a specialized role. Most of the hydroids provide nutrition, some have defensive roles and others produce gametes. Hydroids reproduce sexually or asexually. During typical asexual reproduction small medusae are produced by specialized hydroid polyps. The medusae may either remain attached to the colony or become pelagic. Medusae are usually dioecious and reproduce sexually. Fertilization may occur externally or internally. There are many different life cycles. Typically, the embryo develops into a free swimming larva called a planula that will eventually settle and transform into a polyp. Formation of medusae from polyps usually occurs during certain times of the year and is possibly governed by environmental and hormonal factors.

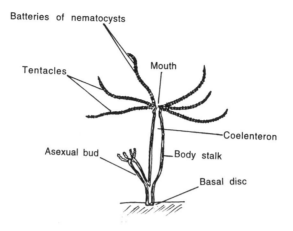

Section through a hydra

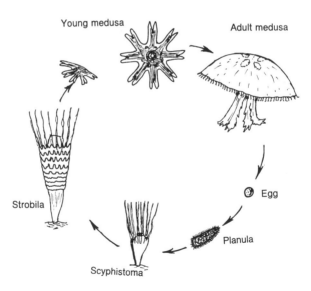

Life cycle of *Aurelia*

Class Scyphozoa (jelly fish)

This is the smallest class of the cnidarians containing the largest medusae. The dominant and more conspicuous body form of scyphozoa is the medusoid form. Even though the medusae of Scyphozoa are generally larger than those of Hydrozoa, they are very similar in appearance. Bathers are familiar with several members of this class because many species occur in coastal waters and possess stinging nematocysts. Most scyphozoans are predators, feeding on zooplankton and fish larvae. Some species capture their prey using toxic nematocysts and others trap plankton on a mucous covered surface. Reproduction is similar to that of hydrozoans. However, in the scyphozoan medusa form, the gametes are usually attached to the gastrodermis, whereas in hydrozoans the gametes are attached to the epidermis. In Scyphozoa, medusae are produced from the polyploids in a similar way as in Hydrozoa. In addition, some scyphozoans form medusae from the polyploid form by transverse fission of the oral end (strobilation).

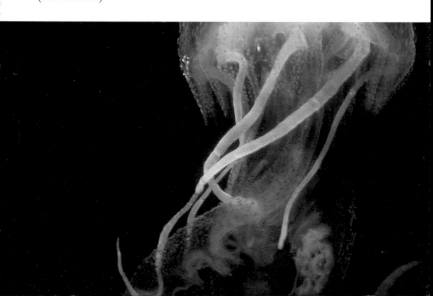

Pelagia noctiluca

<u>Common names</u>: luminescent jellyfish (En), tsuhtra (Gr), leucht-
qualle (G)

<u>Description/Biology</u>: *P. noctiluca* glows at night, hence the name.
This jellyfish has fairly long tentacles and a purple-pinkish
colouration. Its sting is one of the strongest of all jellyfish in
the Mediterranean. For this reason, bathers often avoid
swimming in areas where luminescent jellyfish are abundant.
They prey during the day and at night on a variety of small
animals, particularly crustaceans. They swim by pulsing the
subumbrella and driving water out beneath it. The bell usually
grows to a diameter of 9 cm. Luminescent jellyfish are usually
found between the surface and 10 m depth.

<u>Classification:</u> Order Semaeostomeae, Family Pelagiidae

<u>Comments</u>: The stings from this jellyfish are very painful. Appli-
cation of ammonia or urine can help alleviate pain. The
majority of stinging cells are located on the tentacles, not on
the bell.

Aurelia aurita

<u>Common names</u>: jellyfish (En), yiali (Gr)

<u>Description/Biology</u>: This clear jellyfish has very short tentacles that do not sting. It has both branched radial canals and a ring canal. The bell can reach 40 cm in diameter. Strobilation takes place in winter and large individuals are usually present in the summer.

<u>Classification</u>: Order Semaeostomeae, Family Ulmaridae

<u>Comments</u>: This is a common cosmopolitan species that can often be very abundant.

 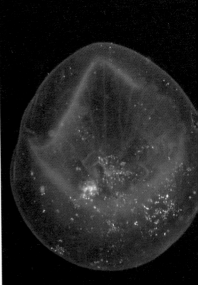

Class Anthozoa (corals, sea fans, sea anemones)

Anthozoa is the largest class of cnidarians, containing over 6,000 species. They can be either solitary or colonial. This class differs from the other cnidarian classes, by the complete absence of the medusoid form. The polyploid form is different than that of the other cnidarians. The mouth is connected to the gut by a pharynx and the gastrovascular cavity is composed of several longitudinal segments that are separated by septa or mesenteries. The nematocysts of anthozoans are present not only on the tentacles but also on the septa. This class can be further separated into 3 main groups: the sea anemones, the stony corals, and the octocorallian corals.

Octacorallian Corals (sea fans, pipe coral, sea pens)

Octacorals are colonial cnidarians. They differ from other anthozoans because they always possess eight septa and eight pinnate tentacles. Unlike the stony corals which possess an external skeleton, the octacorals have an internal skeleton made up of either loose or fused calcareous or horny spicules. Octacorals feed mainly on zooplankton. Many species are hermaphroditic and can reproduce sexually and asexually. This group includes the precious red coral that can be found in the Mediterranean and in Japanese waters.

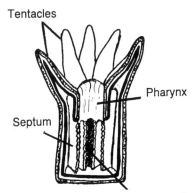

Tentacles

Pharynx

Septum

Gastrovascular cavity

Eunicella cavolinii

Common names: yellow sea fan (En), corali (Gr), gelbe gorgonie (G)

Biology/Description: The yellow sea fan attaches to hard substrates
and forms colonies that are erect and fan like. The profusely
branching fan usually faces into the current and feeds at night
on plankton. It is found between 10 and 100 m depth and may
reach over 30 cm in height.

Classification : Order Gorgonacea, Family Plexauridae

Eunicella singularis

Common names: white sea fan (En), corali (Gr), weibe gorgonie (G)

Biology/Description: This colonial sea fan inhabits waters between 5
and 40 m and may grow to be 30 cm in height. It attaches to
hard substrates and has a preference for unshaded areas. It
feeds at night on zooplankton. The white sea fan does not
branch as much as the yellow sea fan.

Classification: Order Gorgonacea, Family Plexauridae

Sea Anemones

Sea anemones are solitary polyps, larger than other anthozoans. They are often brightly coloured and are usually attached to hard substrates or partially buried in soft bottoms. Although sea anemones are considered sessile, several species can move along the bottom, and some can even swim short distances; some species are even planktonic. They are abundant in coastal tropical waters and can also be found in the temperate zone and in deep waters. A few species are symbiotic with other invertebrates such as hermit crabs, shrimps and fishes. A sea anemone posses a pedal disk, a column and an oral disk. The pedal disk attaches to the substrate. The column usually forms the major part of the body and contains the gastrovascular cavity. The oral disk contains the mouth which is surrounded by eight to hundreds of tentacles. Many sea anemones are able to cover their oral disk by extending and pulling their column over it. Sea anemones usually feed by catching prey (such as invertebrates and fishes) with their stinging tentacles. Adhesive cells, called spirocysts, are used to capture hard shelled prey that may be resistant to nematocysts. Most sea anemones are hermaphroditic and can produce sexually and asexually.

Actinia equina

<u>Common names:</u> beadlet anemone (En), anemoni (Gr), purpurrose (G)

<u>Description/Biology:</u> The beadlet anemone is red to light purple. It lives on hard substrates in very shallow waters from the surface to 0.5 m. Because it lives in the intertidal zone, it may sometimes be exposed to air. This species is nocturnal. During the day it is usually contracted and at night it exposes its tentacles. The oral disc may reach a diameter of 5 cm. Like the other anemones, this species is carnivorous.

<u>Classification:</u> Order Actiniaria, Order Nynantheae, Family Actiniidae

Anemonia rustica

<u>Common names</u>: opelet anemone (En), anemoni (Gr), wachsrose (G)

<u>Description/Biology</u>: The opelet anemone is very similar to its congeneric A. sulcata. It's tentacles are light brown/yellow without red tips that are characteristic of *A. sulcata*. Members of this species can be found living in groups or by themselves on soft and hard substrates at depths from 0.2 m to 20 m. The oral disc, including the tentacles, can reach 20 cm in diameter. This sea anemone preys upon small invertebrates and fish both during the day and at night. Several organisms, such as a gobiid fish *(Gobius buchichii)*, live in association with this species.

<u>Classification</u>: Order Actiniaria, Suborder Nynantheae, Family Actiniidae

<u>Comments</u>: Although they are not usually offered in restaurants or sold at markets, these sea anemones are sometimes eaten fried or are used as an ingredient for anemone balls (similar to meatballs).

Anemonia sulcata

<u>Common names</u>: red tipped anemone (En), anemoni (Gr), wachsrose (G)

<u>Description/Biology</u>: This sea anemone has red tipped tentacles. The red colouration comes from symbiotic algae. Members of this species can be found living in groups or alone on soft and hard substrates at depths from 0.2 m to 20 m. The oral disc including the tentacles can reach 20 cm in diameter. This sea anemone preys upon small invertebrates and fish both during the day and night. Several organisms, such as a gobiid fish *(Gobius buchichii)* live in association with this species.

<u>Classification</u>: Order Actiniaria, Suborder Nynantheae, Family Actiniidae

<u>Comments:</u> Although they are not usually offered in restaurants or sold at markets, these sea anemones are sometimes eaten fried or are used as an ingredient for anemone balls (similar to meatballs).

Aiptasia mutabilis

<u>Common names:</u> trumpet anemone (En), anemoni (Gr), siebane-
mone (G)

<u>Description/Biology</u>: The trumpet anemone has two morphs. One has
partly transparent tentacles and is fairly solitary, and the other
has brownish tentacles and is usually found in close proximity
to conspecifics. The transparent tentacle morph can be found
in depths from 2 m to 20 m. The brown tentacle morph
inhabits shallower waters from 0.5 m to 20 m depth. Both are
active during the day and night and are found on hard
substrates. They are carnivorous and catch their prey with
specialized nematocysts. Trumpet anemones can grow to 10
cm in oral disc diameter. The oral disc includes the tentacles
which contract when the anemone is shaded or threatened.

<u>Classification:</u> Order Actiniaria, Suborder Nynantheae, Family Aipta-
siidae

<u>Comments:</u> Although they are not usually offered in restaurants or
sold at markets, these sea anemones are sometimes eaten fried
or are used as an ingredient for anemone balls (similar to
meatballs).

Calliactis parasitica

<u>Common names:</u> hermit anemone (En), anemoni tou skatsounari/
karkinari (Gr), schmarotzerrose (G)

<u>Description/Biology:</u> This sea anemone is symbiotic with hermit
crabs and gastropods. It attaches itself to the shells of its
symbiont. When disturbed, it emits sticky threads. The hermit
anemone is carnivorous and active both during the day and
night. The oral disc diameter may reach 6 cm. This species is
found on soft and hard substrates, depending on the
symbiont, at depths from 1 m to 30 m. It is cream to light
brown with reddish or grayish linear blotches.

<u>Classification:</u> Order Actiniaria, Suborder Nynantheae, Family
Hormathiidae

Phylum Cnidaria

<u>Stony or Scleractinian Corals:</u> Stony corals contribute greatly to coral reef formations. They are sessile and usually colonial. They produce a calcium carbonate skeleton which is usually made up of a cup and thin septa. The polyps sit on top of the cup and secrete calcium carbonate that is deposited below them. The calcium carbonate skeleton provides both structure for the polyps to live on and protection from predators. When threatened, the polyps are able to withdraw into the skeleton. Many corals are nocturnal; they feed at night and withdraw into their skeleton during the day. Stony corals prey on organisms ranging from zooplankton to small fishes. Once captured, the prey is stung by the polyp's nematocyst-laden tentacles. The majority of stony corals also obtain nutrients from symbiotic zooxanthellae. Many species are hermaphroditic and can reproduce sexually and asexually. Occasionally corals compete for space, in such situations aggressive species secrete digestive enzymes onto their neighbors and digest the tissue.

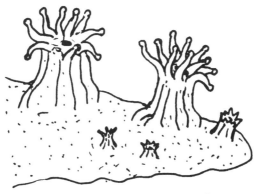

Extratentacular budding of a coral

Madracis sp.

<u>Common names:</u> coral (En), corali (Gr), korallen (G)

<u>Description/Biology:</u> This coral forms encrusting, nodular or branched colonies. The size of the branches depends on the species, but usually the width does not exceed 1.5 cm. They can have green, red, yellow, brown or cream colour. The polyps can usually be seen (at least partially) during the day. Most of the species have 10 well developed septa with smooth margins.

<u>Classification:</u> Order Scleractinia, Family Pocilloporidae

<u>Comments:</u> Madracis species are fairly common and occur on most coral reefs.

Eusmilia fastigiata

<u>Common names:</u> coral (En), corali (Gr), korallen (G)

<u>Biology/Description:</u> Although *Eusmilia* forms colonies (around 50 cm in diameter) of well separated corallites, individual corallites are often seen. *E. fastigiata* has numerous leafy septa which are prominent and have smooth margins. The oral disks are usually about 1.5 cm in diameter. During the day it is difficult to distinguish the polyps and tentacles because they are retracted into the skeleton. They have a yellowish, brown or bluish colour with colourless tentacles. This coral can be found in a variety of habitats in both shallow and deep waters.

<u>Classification:</u> Order Scleractinia, Family Caryophylliidae

Balanophyllia (Leptopsammia) pruvoti

Common names: yellow cup coral (En), kitrino corali (Gr), gelbe nelkenkoralle (G)

Biology/Description: This small solitary coral has a bright yellow polyp. It can be found on hard substrates, at depths between 1 m and 50 m. It is nocturnal and feeds on zooplankton. The oral disc including the tentacles may reach a diameter of 2 cm.

Classification: Order Scleractinia, Dendrophyllidae

Comments: Even though this is a very colourful coral, because of its size and nocturnal behaviour, it usually goes unnoticed by divers.

Cladocora caespitosa

<u>Common names:</u> sand coral (En), tragana (Gr), rasenkoralle (G)

<u>Biology/Description:</u> Sand coral is a colonial stony coral usually
found on sandy or soft substrates in shallow turbid waters.
The colonies are small and never assume the character of a
reef. The brownish colonies posses narrow elongated
corallites that arise from an encrusting base.

<u>Classification:</u> Order Scleractinia, Family Faviidae

Phylum Ctenophora (comb jellies)

This is a small marine phylum that consists of approximately 50 species. Ctenophores are mostly transparent planktonic organisms that may look similar to jellyfish; however, with the exception of one species, ctenophores do not have nematocysts. The body is divided by eight comb rows that are made up of cilia. The beating comb rows provide propulsion for the ctenophore. Some species possess tentacles that contain adhesive cells that trap zooplankton. Other species lack tentacles and may prey upon other ctenophores by engulfing them. Ctenophores are luminescent; they can emit flashes of light. All species are synchronous hermaphrodites. The gametes are contained in canals within the body. Fertilization is usually external; the eggs develop into planktonic larva which metamorphosize into adults. These gelatinous creatures are very delicate and difficult to collect undamaged with nets.

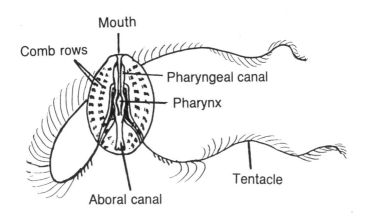

A comb jelly (*Pleurobranchia*)

Bolinopsis infundibulum

Common names: comb-jelly (En), mixa (Gr), rippenquallen (G)

Biology/Description: This comb-jelly has fairly small tentacles with adhesive cells that trap zooplankton. There are two large lobes on either side of the mouth. Like all comb-jellies, swimming is achieved by the beating of the eight comb rows that are iridescent in light. They are active during the day and night. This species can grow up to 8 cm in length.

Classification: Class Tentaculata, Order Lobata, Family Bolinopsidae

Comments: Comb-jellies are transparent and are thus usually difficult to see; however, at night they can bioluminesce and emit a greenish flash of light. A related genus *(Mnemiopsis)* is thought to be responsible for the decline in the Black Sea fisheries because larval fishes compose a major part of its diet.

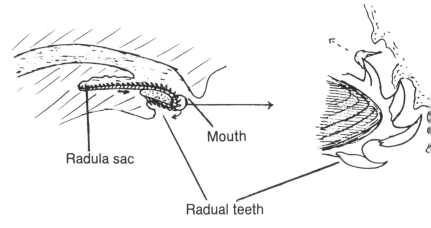

Mollusca radula scraping against the substratum

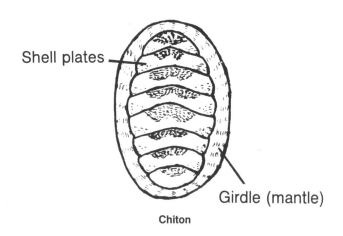

Chiton

Phylum Mollusca (molluscs)

Mollusca is the second largest invertebrate phylum. Marine molluscs can be found at all depths. Because they are so abundant, most of the shells that are found on beaches are produced by molluscs. Most molluscs are characterized by the presence of a radula (a toothed tongue that is used for feeding) and a mantle. In some species, the body wall folds to form a mantle. In others, it extends to form tentacles, or thickens to form a foot. The mantle may secrete a calcareous shell or it may be used for protection or locomotion. Gills are usually enclosed in the mantle cavity. Many molluscs live part of their life as planktonic larvae called veligers.

Since prehistoric times molluscs have been used as a food resource for humans. In later periods, they were also used as currency and for dye. Currently, they are valued because they provide both food and beautiful shells.

Class Polyplacophora (chitons)

Chitons have an oval body shape that is dorsoventrally flattened and is covered by a shell consisting of eight interlocked transverse plates. The shell is attached to a mantle which forms a toughened girdle around the periphery of the shell. The girdle is often fringed with spines, calcified granules or small plates on the dorsal side. The hard shell provides protection from predators and the interlocking plates increase flexibility allowing chitons to attach to curved rocky surfaces. They have a small head with no eyes or tentacles. The low profile reduces water resistance and the large flat foot that occupies most of the ventral surface help in locomotion and adhesion in turbulent environments. There are approximately 800 species of Polyplacophora, which can usually be identified by the morphological characteristics of their plates and girdle. Most species are between 3 and 12 cm in length and are dull yellow, brown, green or red. Many are intertidal and are found on hard substrates where they feed on algae which they scrape with their long radula that bears 17 teeth in each transverse row. Most prefer dark areas and are more active at night. The sexes are separate and fertilization is external. In some species the female broods the eggs in the mantle cavity. The embryos of non brooding species pass part of their lives as free swimming trochophore larvae.

Chiton olivaceus

<u>Common names:</u> chiton (En), hitonas (Gr), Käferschnecke (G)

<u>Description/Biology:</u> This chiton has an oval shape that is not very elongate. It has teeth on its partially exposed plates. The colouration can vary from different shades of yellow, brown, olive or red. It lives on hard substrates at depths from approximately 0.5 m down to 5 m, and feeds mostly at night, scraping algae using its long radula. It is usually around 2 to 3 cm and can grow up to 4 cm in length.

<u>Classification:</u> Order Ischnochitonina, Family Chitonidae

<u>Comments:</u> When dislodged it curls up into a ball. This behaviour serves for protection and allows the chiton to upright itself. Chiton is the only genus from the family Chitonidae represented in the Mediterranean.

Class Gastropoda (snails, limpets, abalone)

The gastropods are the largest and most diverse group of molluscs. The majority of them possess a single calcareous shell, which may be spiral or flattened. The shape and colour of this shell is important in species identification. Gastropods have a clearly defined head with sense organs (tentacles and eyes) and a mouth with either a radula or jaws. The foot is a defined flat structure used for locomotion. The mantle secretes a calcareous shell and often includes a siphon for respiration. In some species such as cowries, the mantle covers the calcareous shell. Species belonging to this group can be either carnivorous or herbivorous. Certain carnivorous gastropods (e.g. the cone shells) produce poison that is used to incapacitate prey and ward off predators. Gastropods are synchronous hermaphrodites, an individual can posses both male and females gonads at the same time. When two organisms meet, each one can simultaneously act as a male and female. They usually lay their eggs in clutches on the substrate. In some species the hatching young are in the form of the adult (e.g. whelks); in other species the young spend part of their lives as planktonic larvae.

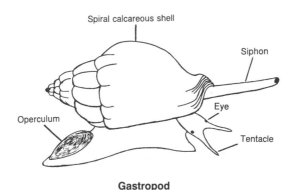

Gastropod

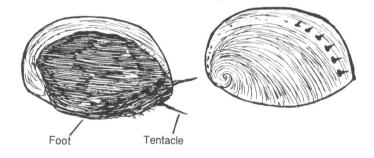

Foot   Tentacle

Dorsal and ventral view of *Haliotis*

Haliotis lamellosa

Common names: abalone (En), haliotida or afti tis thalassas (Gr),
seeohr (G)

Description/Biology: Haliotis lamellosa is smaller (usually 4 to 6 cm,
can reach up to 10 cm in length) than its Pacific counterparts
(30 cm in length); this may be one of the reasons why they are
not targeted heavily by Greek fishermen. Their shell is
flattened and possesses a series of 7 to 8 holes along the outer
edge. These holes are used for excretion and respiration. The
internal surface of the shell has a bright iridescent silvery
colour while the external surface is dull green-brown and is
usually covered by algae and bryozoans. Young individuals
may have a brown, white or red external shell colour.
Abalones move and cling to rocky surfaces using a powerful
foot and feed on algae attached to the rocky surfaces. They
can be found in rocky environments from 1 to 10 m depth.

Classification: Order Archaeogastropoda, Family Haliotidae

Comments: Known for their beautiful shells and tasty meat, abalones
have been heavily exploited in the Pacific. They are usually
eaten broiled. Even though they are very common along
Mediterranean coastlines, they are not found in Greek
seafood markets. Abalones can also be used as fish bait. Some
scientists separate *Haliotis lamellosa* into two species, *H.
lamellosa* and *H. tubercolata*.

Patella caerulea

<u>Common names:</u> limpet (En), petalida (Gr), napfschnecke (G)

<u>Description/Biology:</u> Limpets have a flattened conical shell and a powerful foot that is used for locomotion and secures them very firmly to rocks. These organisms return to the same site after feeding expeditions. Herbivorous, limpets graze on algae that grow on rocky surfaces. They grow up to 5 cm in length and live on rocks from 0 to 8 m depth.

<u>Classification:</u> Order Archaeogastropoda, Family Patellidae

<u>Comments:</u> They are edible and can be eaten raw or broiled. Even though they are not found in Greek seafood markets, they are common in the Mediterranean. Limpets make good fish bait.

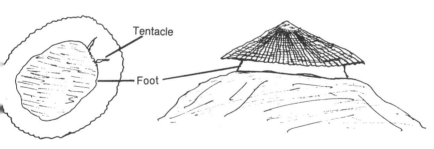

Tentacle

Foot

Patella

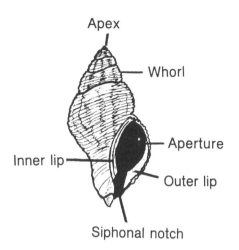

Apex

Whorl

Aperture

Inner lip

Outer lip

Siphonal notch

Gastropod shell

Gibbula magus

<u>Common names:</u> turban top shell (En), salingaraki (Gr), würfelturban (G)

<u>Description/Biology:</u> Turban top shells have an operculum and a conical shell. The profile of the shell is markedly stepped with a prominent outer keel of the last whorl. Each whorl has short nodular ribs on a number of spiral ridges and grooves. The shell is usually white, grey or yellow and has red, purple or brown blotches. They feed on detritus and grow to around 3 cm in diameter. Turban top shells can be found at depths down to 70 m on a variety of different substrates. *G. magus* is the largest species of *Gibbula* in the Mediterranean.

<u>Classification:</u> Order Archaeogastropoda, Family Trochidae

<u>Comments:</u> This is a fairly common species in the Mediterranean and is often used as fish bait.

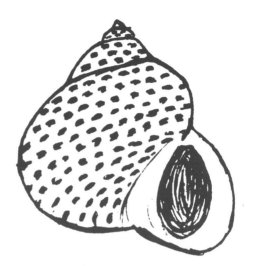

Monondata

Monondata articulata

Common names: turban shell (En), salingaraki (Gr), würfelturban (G)

Description/Biology: This gastropod looks very similar to *Monondata turbinata*, but it has finer grain markings that follow a "zig-zag" pattern and a longer shell. Turban shells are protected by a thick calcareous spiral shell. When threatened by predators or dehydration these organisms can retreat into their shells and close the entrance with a tough horny operculum. These small gastropods have a diameter of 3 cm and length of 4 cm. They are herbivorous and scrape algae from rocky surfaces. These organisms are usually found in rocky environments in shallow waters, less than 2 meters deep, and in tide pools.

Classification: Order Mesogastropoda, Family Trochidae

Comments: They are common in most of the Mediterranean and are often used as fish bait.

Monondata turbinata

<u>Common names:</u> turban shell (En), salingaraki (Gr), würfelturban (G)

<u>Biology/Description:</u> This turban shell has thicker black markings than *M. articulata.* The markings are on a yellowish-brown background that may sometimes appear green because of algae on the shell. When threatened by predators or dehydration, turban shells can retreat into their thick calcareous spiral shells and close the entrance with a tough horny operculum. *M. turbinata* attains a diameter of 2.5 cm and length of 3 cm. They are herbivorous and scrape algae from rocky surfaces in shallow waters (usually less than 2 meters) and tide pools.

<u>Classification:</u> Order Mesogastropoda, Family Trochidae

<u>Comments:</u> They are very common in most of the Mediterranean and are often used as fish bait.

Tricolia speciosa

<u>Common names:</u> pheasant shell (En), cohilaki (Gr)

<u>Description/Biology:</u> Pheasant shells have a turban or conical shaped glossy shell, with the last whorl approximately two thirds of the shell length. The colour of the shell varies from white to red to dark brown. It is rarely one colour, but rather is covered in dark red, brown or purple streaks and blotches that may sometimes be in a "zig zag" pattern. The obviously convex operculum is white. These herbivorous gastropods are small in size (up to 0.6 cm in diameter and 1.4 cm in length) and live in rocky environments from the littoral zone to 40 m depth.

<u>Classification:</u> Order Archaeogastropoda, Family Tricoliidae

<u>Comments:</u> Because they are fairly common in the Mediterranean, empty shells of these gastropods can often be found on sandy beaches. There are only five species of *Tricolia* in the Mediterranean.

Trivia multilirata

Common names: cowrie (En), cohilaki (Gr), kauri (G)

Description/Biology: The cowrie shell is egg shaped and has strong
ridges and groves (to a certain extent,these have been eroded
in the specimen photographed). The shell is flattened on the
aperture side, and the narrow ridged aperture runs the whole
length of the shell. The mantle extends laterally and almost
entirely covers the shell. Small individuals do not have
prominent ridges and have a much wider aperture.
Individuals grow to 1.2 cm length and 0.8 cm in width. The
shell is pinkish white. *T. multilirata* usually inhabits rocky
environments and feeds on ascidians.

Classification: Order Mesogastropoda, Family Eratoidae

Comments: Sometimes empty cowrie shells are found on sandy
beaches.

Semicassis undulata

Common names: little ton (En), kohili (Gr)

Description/Biology: This gastropod has a thick, spirally grooved
shell that is brownish yellow in colour with varying amounts of
brownish red markings. The shell may grow up to 10 cm. Both
lips are thick and the outer lip has tooth-like grooves. The
little ton inhabits soft bottoms in the sublittoral zone where it
preys on echinoderms.

Classification: Order Mesogastropoda, Family Cassidae

Comments: Empty shells are nice collectible items. It is fairly
common in the Mediterranean.

Tonna galea

<u>Common names:</u> giant ton (En), kohila (Gr), tonnen (G)

<u>Description/Biology:</u> This gastropod has a relatively thin light brown
grooved shell and lacks an operculum. The outer lip is thin
and there is no visible inner lip. The shell grows to about 20
cm in diameter. The giant ton is carnivorous and preys at
night on other molluscs and echinoderms. *T. galea* uses
secretions of sulfuric acid to penetrate their hard protective
shells. It inhabits areas between 2 m and 100 m depth. During
the day it stays hidden in the sand or in *Posidonia* beds.

<u>Classification:</u> Order Mesogastropoda, Family Tonnidae

<u>Comments:</u> Even though this species is not really found in Greek
seafood markets, it is fairly common in fish markets in Sicily.
As it lives in fairly deep waters, it is usually only brought to the
surface in fishermen's nets.

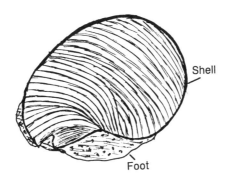

Shell

Foot

Tonna galea

Murex brandaris

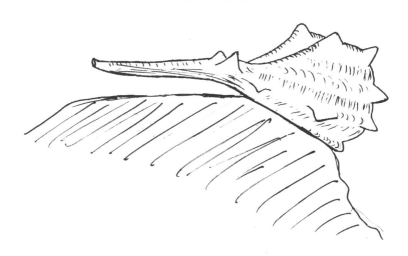

Eggs of *Murex sp.*

Murex brandaris

Common names: murex (En), porfira (Gr), herkuleskeule (G)

Description/Biology: Murex has a hard protective yellowish pink shell with several spines. It is most active at night and inhabits sandy habitats. This carnivorous gastropod preys on both live and dead organisms, including other molluscs. Some species of this genus are adapted for drilling holes in hard shelled prey, such as limpets, barnacles and especially bivalves. Other species can pull or wedge the shells of bivalves apart. Using its long siphon, it is able to detect chemical signals from its prey. It usually grows up to 9 cm in length and lives in the littoral and sublittoral zones on soft substrates. Sometimes murex can be found in large aggregations.

Classification: Order Neogastropoda, Family Muricidae

Comments: Even though it is not really found in Greek seafood markets, it is fairly common in Italian fish markets. Murex is usually eaten boiled or broiled and can be used as fish bait. In ancient times murex were used as a source of expensive purple dye.

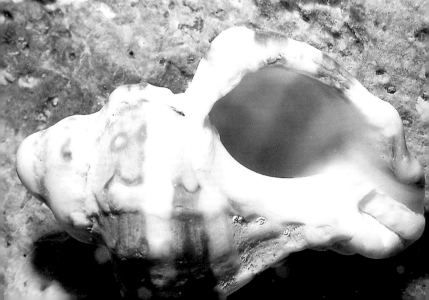

Trunculariopsis trunculus

Common names: murex (En), porfira (Gr), herkuleskeule (G)

Description/Biology: The hard protective shell has several spines, shorter than those of *Murex. brandaris*. They live in sandy habitats and are most active at night. These carnivorous gastropods prey on both live and dead organisms; their prey including other molluscs. Some species of this genus are adapted for drilling holes in hard shelled prey, such as limpets, barnacles and especially bivalves. Other species can pull or wedge the shells of bivalves apart. They are able to detect chemical signals from their prey using their long siphon. In one of the pictures shown, the shell is covered in algae and in the other the spines have been eroded (this specimen belongs to the subspecies *Trunculariopsis trunculus vericosus*). The shell can reach sizes from 4 cm to 13 cm in length. *T. trunculus* usually inhabit rocky areas in the littoral and sublittoral zones.

Classification: Order Neogastropoda, Family Muricidae

Comments: Even though *T. trunculus* is not found in Greek seafood markets, it is fairly common in Italian fish markets. *Murex* is usually eaten boiled or broiled and can be used as fish bait. These gastropods often get tangled while feeding on fish caught in fishermen's nets and so they are often brought up with the catch. This species as well as Murex brandaris were used in ancient times as a source of die. Other species within the family Muricidae are sought after by collectors as they have intricate shapes and long spines. This species is very common in the Mediterranean.

Conus having captured a goby

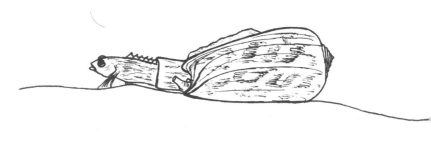

Conus mediterraneus

Common names: cone shell (En), cohili (Gr), schnecken (G)

Description/Biology: *C. mediterraneus* is probably the only species of
cone shell in the Mediterranean. The family Conidae has
many representatives in tropical and subtropical seas such as
the Western Atlantic and Indo-Pacific oceans. Their shells
are cone shaped, hence the name. The colour of the shell
varies; however, it usually has shades of green and brown. The
aperture is straight and long. The shell can grow up to 7 cm in
length and up to 4 cm in diameter. Cone shells live in the
littoral and neritic zone. They are carnivorous, feeding
primarily on polychaete worms, other gastropods or fish.
They catch their prey using singular poisonous, barbed,
harpoon type radular teeth. The spears are thrust into their
prey with the use of the cone's long and highly maneuverable
proboscis. Species that prey on fish and polychaetes hide in
the sand and strike when the prey pauses above them. The
harpoon is thrust into the soft belly of the victim and stays
connected to the cone shell. In species that prey on other
gastropods, the harpoon is not attached to the cone after the
strike.

Classification: Order Neogastropoda, Family Conidae

Comments: *C. mediterraneus* is fairly common in the Mediterranean.
Some species in the family Conidae from the South Pacific
have highly toxic "harpoons" that are able to kill a human
within 4 hours. These species should be handled with the
utmost of care.

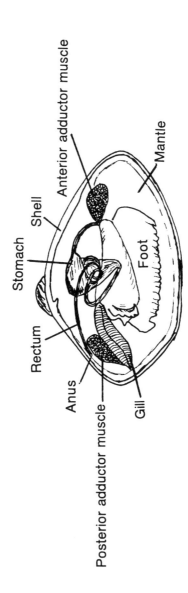

Body of a bivalve with the right valve removed

Anterior adductor muscle

Mantle

Shell

Stomach

Foot

Rectum

Anus

Posterior adductor muscle

Gill

Class Bivalvia or Pelecypoda (mussels, oysters, clams)

These molluscs can be distinguished by two hard, usually symmetrical, shells protecting a soft body. An elastic ligament keeps the hinged shells apart during feeding and resting. The hinge is formed by interlocking tooth-like structures. These structures as well as the shape and colour of the shells are useful taxonomic characters. One or two strong adductor muscles close the two shells firmly when the organism is under attack. Bivalves have also a characteristic muscular hatchet shaped foot. Some bivalves, such as razor clams, use this well developed foot to burrow in soft substrate. Other non burrowing bivalves, such as scallops, mussels and oysters, have an atrophic almost non-existent foot. Burrowing bivalves posses a single or double siphon which is used for respiration and feeding. All bivalves feed on phytoplankton by filtering large volumes of water through their gills. These organisms reproduce sexually. Males and females release gametes into the water column. The fertilized eggs eventually hatch into larvae which spend part of their lives as planktonic organisms before they take the adult form and become sedentary. Small particles, such as sand, that get trapped inside bivalves are sometimes coated with a hard nacreous shell, producing pearls. Usually the pearls remain small, 1-3 mm in diameter. Certain species found in the Indo-Pacific produce large marketable pearls.

Mytilus edulis

<u>Common names:</u> mussel (En), midi (Gr), miesmuschel (G)

<u>Description/Biology:</u> These organisms are usually found in colonies anchored to structures such as piers, pilings, and rocks. The shell is elongated and symmetrical and may be covered with barnacles, algae, and bryozoans. The colour of the shell can vary from purplish brown to blueish black. Occasionally the shell has a yellowish brown colour with orange brown lines. Mussels attach themselves to each other and to structures with byssal threads. Although they appear as sessile organisms, they have the ability to release their byssus threads and move over short distances using their small foot. These bivalves filter large quantities of water and have a high tolerance for pollutants and eutrophic conditions. In fact, high concentrations of mussels are often found close to harbors and other polluted areas. Because mussels can survive in very polluted waters and bioaccumulate heavy metals in their tissues, they are used as bioindicators and for pollution studies. *M. edulis* can grow to be up to 16 cm, however the usual size is around 9 cm. Another common species is *M. galloprovincialis* which looks very similar to *M. edulis* but has a broader less elongated shell. Mussels are usually found at depths less than 5 m.

<u>Classification:</u> Order Filibranchia, Family Mytilidae

<u>Comments:</u> Mussels are a delicious food item that can be cooked in a variety of ways. Because they bioaccumulate heavy metals, it is not safe to eat mussels from polluted areas. However, mussels found in Greek markets are usually from aquaculture plants and are therefore presumed safe for consumption. This is an Atlantic species that is not very common in the Eastern Mediterranean.

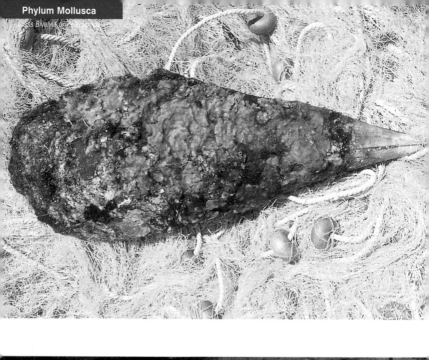

Pinna nobilis

Common names: pen shell (En), pina (Gr), steckmuscheln (G)

Description/Biology: Pen shells have two symmetrical teardrop-shaped shells. The external surfaces of the valves have numerous sharp spine like projections that can be very abrasive. The shells are yellowish brown externally and pearl - brown coloured internally. Pen shells are sessile; they anchor themselves in soft substrate using byssus threads. At least one forth of the shell is embedded in the substrate. These bivalves are the largest that are found in the Mediterranean. They can grow up to 1 m long and are found between 5 m and 50 m depth on sandy bottoms close to *Zostera* or *Posidonia* beds. Pen shells are not very common. They prefer environments with low turbidity, slight current and low annual salinity change. Several organisms such as symbiotic crabs and shrimp often live inside pen shells.

Classification: Order Filibranchia, Family Pinnidae

Comments: Pen shells can be eaten fried. However, they need special preparation in order to remove the slight "charcoal" taste. They are not usually found in Greek seafood markets. Small shells are very fragile and transparent.

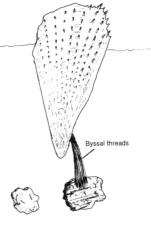

Byssal threads

Pina

Ostrea edulis

<u>Common names:</u> oyster (En), stridi (Gr)

<u>Description/Biology:</u> Oysters have a thick shell that tapers toward the edges, becoming thin and brittle. One of the unequal valves is flat, while the other (the larger one) is convex. The convex valve is attached to either a rock or other hard substrate such as other oysters, man-made structures, etc. Oysters are often found together in large numbers, forming oyster beds. The flat valve is brownish-gray, the convex valve is pale brown, greenish or yellowish. Inside the valves are pearly white. *O. edulis* can grow up to 12 cm and can be found at depths down to 100 m.

<u>Classification:</u> Order Eulamellibranchia, Family Ostreidae

<u>Comments:</u> Oysters are usually eaten raw and sometimes smoked, with ouzo or white wine. It is important that the oysters are consumed fresh to prevent fatal food poisoning. Aquaculture of this species has been developed in order to satisfy the market demand.

Chlamys pesfelis

<u>Common names:</u> scallop (En), hteni (Gr), pilgermuschel (G)

<u>Description/Biology:</u> Scallops are one of the most mobile group of bivalves. Most scallop species are able to move by expelling water from inside their shells by rapidly snapping their valves. However, some species live attached to hard substrate with the help of byssal threads. Both valves of *C. pesfelis* are ribbed and convex. The wings of the shells are uneven in size. Internally they are pinkish white coloured, and externally they are brownish. A row of eyes and tentacles along their mantle edge alerts these bivalves of approaching predators. They grow to 12 cm and are usually found on sandy and muddy bottoms at depths between 3 m and 50 m.

<u>Classification:</u> Order Filibranchia, Family Pectinidae

<u>Comments:</u> Scallops are considered a delicacy and accompany ouzo

well. They are usually eaten broiled, fried or raw. *C. pesfelis* together with the smaller more common bay scallop, *Aequipecten* sp., are usually sold in Greek fish markets. Although, the larger scallop, *Pecten jacobaeus*, is rarely found in Greek markets, it is cultured in other parts of Europe. *C. pesfelis* is fairly common in the Mediterranean.

Callista chione

<u>Common names:</u> cockle (En), ahivada (Gr)

<u>Description/Biology:</u> These bivalves have a broad ovate and moderately convex shell. They have smooth and glossy valves that are usually reddish brown or pinkish brown with darker rays. The inside of the shell is whitish. Cockles inhabit areas with soft substrate within the littoral and sublittoral zones. They grow up to 9 cm and usually live in water from 4 m down to 215 m.

<u>Classification:</u> Order Eulamellibranchia, Family Veneridae

<u>Comments:</u> They make a good appetizer (mezedaki), and they can be eaten raw and possibly broiled. Although they are not found in Greek seafood markets, they are available in Italian markets.

Donacilla cornea

<u>Common names:</u> banded wedge shell (En), mikri ahivada (Gr)

<u>Description/Biology:</u> These small bivalves have a smooth shell that is oval, slightly triangular and elongated. The valves can be many different colours such as yellow, grey, red, brown, white and cream. They are common and can be found buried in the sand on beaches in the littoral zone. Their maximum size is approximately 2.5 cm in length.

<u>Classification:</u> Order Veneroida, Family Mesodesmatidae

<u>Comments:</u> Although they are very common in the Mediterranean, they are easily missed by the casual observer because of their small size and because they are usually berried in the sand.

Ensis ensis

<u>Common names:</u> razor clam (En), solinas (Gr)

<u>Description/Biology:</u> These elongated bivalves have a long thin narrow shell (thus their name) with curved edges that are rounded at the ends. The valves are white and are covered with a yellowish green protective corneous coating known as periostracum. Razor clams bury and anchor themselves in soft substrate using their powerful foot. They are found in sandy or muddy areas, often in colonies. These bivalves may grow to up to 18 cm in length and inhabit waters down to 100 m.

<u>Classification:</u> Order Eulamellibranchia, Family Cultellidae

<u>Comments:</u> They are edible and can be cooked in a variety of ways. Even though razor clams are delicious, they are not usually sold in Greek markets. They are not very common in the Mediterranean.

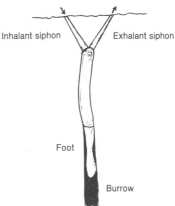

Inhalant siphon   Exhalant siphon

Foot

Burrow

Razor clam, *Ensis*

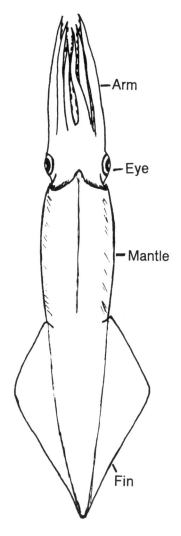

Dorsal view of a squid

Class Cephalopoda (octopus, squid, cuttlefish)

Of all the invertebrates, this group has the most advanced nervous system. It also contains some very intelligent organisms. Cephalopods have a well developed body along the dorsal ventral axis. The head is surrounded by a number of tentacles used mainly in feeding. On the top of the head there are two distinctive well developed eyes. The body of some cephalopods, such as octopus, is encompassed in a tough muscular mantle. In other cephalopods, such as nautilus, the body is protected by an external shell. To move quickly through the water cephalopods have developed a jet propulsion mechanism; water is drawn into the mantle cavity and expelled under pressure through a siphon. The skin contains chromatophores that enable the organisms to quickly change their colour for camouflage and communication. When threatened, many cephalopods may release a cloud of ink through their anus and swim away. All cephalopods are efficient predators, hunting primarily at night. In certain countries, some cephalopod species are considered very important food items. For this reason fishing techniques have been developed to target various cephalopods, and as a result they are heavily exploited.

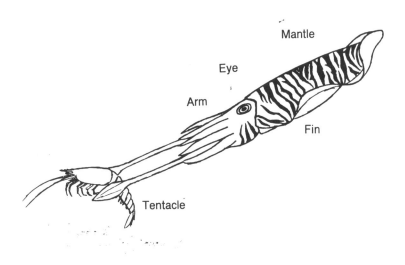

The cuttlefish, *Sepia* capturing a shrimp

Mantle

Eye

Arm

Fin

Tentacle

Sepia officinalis

Common names: cuttlefish (En), soupia (Gr), gemeiner tintenfisch/
sepie (G)

Description/Biology: The cuttlefish has a total of 10 tentacles: 2 long
tentacles are used for prey capture and 8 short ones are used
for holding. Cuttlefish have long narrow fins around the
mantle that are used primarily for stabilization. These
cephalopods are not as sedentary as *Octopus vulgaris*.
Although they are capable of hiding in soft substrate,
cuttlefish are usually found lurking just above the sea floor.
They possess a cartilaginous supporting structure (internal
shell) inside their mantle. Like octopuses, cuttlefish have
poison glands that are used to immobilize their prey. Their
diet consists of crabs and other crustaceans, as well as small
molluscs and other cuttlefishes. Fertilization is internal and
similar to octopi. Eggs are laid at the end of the summer in
clutches attached to the substrate. Cuttlefish grow up to 45 cm
(mantle length) and 4 kg in weight, and can be found in
shallow waters down to about 200 m; however, they are most
abundant in the upper 100 m.

Classification: Order Sepioidea, Family Sepiidae

Comments: Cuttlefish are a fairly popular menu item in Greece, and
they can be cooked in a variety of ways. Although they are not
as common and popular as squid, they can often be found in
Greek seafood markets. Cuttlefish are an important
commercial resource throughout their range.

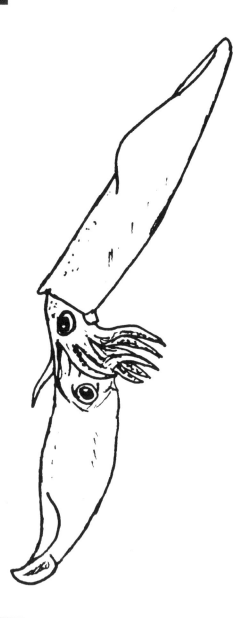

Copulation in the squid, *Loligo*

Loligo vulgaris

Common names: squid (En), calamari (Gr), kalmar (G)

Description/Biology: Squid are the fastest group of cephalopods. Their body is more elongated and torpedo shaped than that of the cuttlefish and their fins are shorter and wider than the fins of cuttlefish. The wider fins move slowly during hovering and much faster during accelerated swimming. Like the cuttlefish, squid have 8 short "holding" tentacles and 2 long "prey capturing" tentacles. At night they move up from deep waters to prey primarily on pelagic fishes closer to the surface. Squid do not posses poison glands. Unlike the octopus that picks the soft parts of its prey and the cuttlefish that chews its food, squid tend to eat their prey whole or piece by piece. Squid feed on fishes and crustaceans, and cannibalism is also common. Inside the mantle there is a cartilaginous supporting structure called a gladius or pen. Squid are pelagic, usually gregarious. Reproduction occurs in large groups. Eggs are left on the substrate in clutches. After mating the majority of adults die. *L. vulgaris* ranges in depth from the surface to 500 m, but is most abundant from 20 to 250 m. Males reach a maximum mantle length of 42 cm and 1.5 kg in weight and females reach a mantle length of 32 cm.

Classification: Order Teuthoidea, Family Loliginidae

Comments: This is a very important commercial species. Squid is a standard dish in many Greek restaurants. Although "calamari" are usually fried, there is a variety of other ways to cook these delicious invertebrates. They are common in Greek seafood markets.

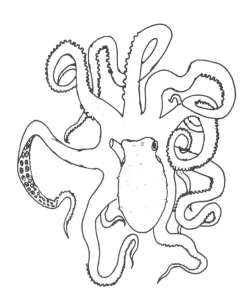

Octopous macropus

Octopus macropus

Common names: white spotted octopus (En), melidona/htapodi (Gr)

Description/Biology: This octopus has conspicuous white spots on a
redish background. The general morphology is similar to that
of *O. vulgaris*, however *O. macropus* has a more slender body
and smaller tentacles. Its life span probably does not exceed
one year. The white spotted octopus is a benthic shallow
water species that can be found in warm temperate waters. It
preys on crustaceans, molluscs and sometimes fishes. It
reaches a maximum total length of 150 cm (maximum mantle
length 14 cm, maximum weight 2 kg), however 60 cm
individuals (total length) are most common.

Classification: Order Octapoda, Family Octopodidae

Comments: This octopus is not very common, thus it is not usually
found in seafood markets.

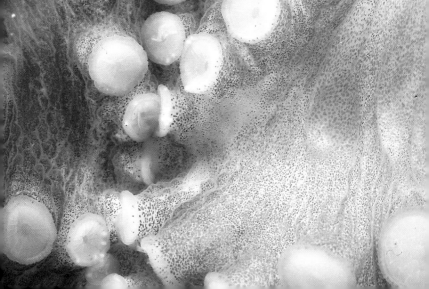

Octopus vulgaris

Common names: octopus (En), htapodi (Gr), gemeiner krake (G)

Description/Biology: The body is muscular without an internal or external shell. There are eight large well developed tentacles surrounding a powerful beak. Two rows of suckers run down the underside of each tentacle. The colouration is usually various shades of brown and gray matching the surrounding environment of the octopus. However, when the octopus dies, it turns light grey. In the winter, larger individuals are found in shallower waters than they are found in during the summer. Usually octopuses inhabit rocky areas, although they can be found over most habitats. In the daytime they stay in a shelter and at night actively forage. Octopuses are very intelligent; for example, they can learn by observing the activity of another octopus. Reproduction occurs in May. Males posses an extra long tentacle that is used to deposit a spermatophore (a mass of sperm) from its own mantle cavity into a female's. Females lay thousands of eggs which they aerate and protect for two to three months. The female only reproduces once in her lifetime; after the young hatch, the female dies. Only a few young survive to become sexually mature. Octopuses feed mostly on crustaceans, molluscs and fishes. They capture their prey using their long tentacles and keep a firm hold with their powerful suckers. The prey is sometimes paralyzed by poison secreted by salivary glands. This octopus can grow up to 1.5 m (total length) and maximum weight 10 Kg. It can be found in shallow waters less than 1 m and to depths greater than 200 m.

Classification: Order Octapoda, Family Octopodidae

<u>Comments:</u> Octopuses are delicious and can be cooked in several ways. Although it is unusual in Greece, octopus can also be eaten raw. Before cooking it is necessary to remove the internal organs and to tenderize the meat. The organs are removed by reversing the mantle and ripping them out. Tenderizing can be done in two ways. The most common procedure is to throw the octopus (at least 40 times) on a smooth hard surface, such as a rock on the beach. Then the octopus is rubbed on a similar surface in a circular movement for several minutes. This will cause a pinkish-white foam to be released. At this point the octopus should be tender, smaller in size, a light gray colour, and the tentacles should be curled. Alternatively, the octopus can be frozen for several months. Octopus is common in Greek seafood markets and is a popular appetizer (mezedaki) in Greek tavernas. Most of the time it is broiled and accompanied with ouzo.

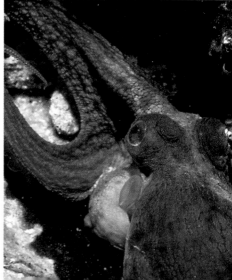

Phylum Nemertea (ribbon worms)

Ribbon worms are unsegmented, bilaterally symmetrical worms with a separate mouth and anus. They are sometimes known as proboscis worms as they have a muscular proboscis apparatus used to capture food. The proboscis can either be branched or armed with a calcified stylet. Prey capture is aided by neurotoxic secretions. Species with a stylet repeatedly stab their victims to increase the effect of the toxin. Species with a branched proboscis wrap their proboscis around their prey and secrete a sticky toxin that immobilizes their victim. The prey can either be swallowed whole or the tissues alone can be sucked into the mouth. There are approximately 900 species in this Phylum, most of which are benthic marine animals. Some species are pelagic. Most of the benthic species live under shells and stones, in algae, or burrow in soft sediments. Some have symbiotic relationships with bivalves, crabs and tunicates. Nemerteans are carnivorous and usually prey on annelids and crustaceans. The majority of species are dioecious and fertilization is external in most of them. The eggs are either deposited in or on the substrate or dispersed in the water column. Certain species reproduce asexually by fragmentation.

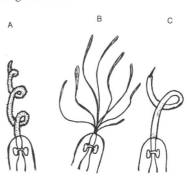

Three different types of nemertean proboscis.
A, proboscis with sticky papillae.
B, branched proboscis.
C, proboscis with calcified stylet.

Cephalothrix linearis

<u>Common names</u>: worm (En), skouliki (Gr), wurm (G)

<u>Description/Biology</u>: The head is somewhat pointed and the mouth is four or five body widths from the anterior tip. When disturbed it contracts into a tight knot. It grows up to 30 cm in length and between 0.05 and 0.1 cm width. The colour is a translucent white, yellowish or grayish. It can be found in the lower shore and sublittoral areas either burrowed in sand or mud, beneath stones, on hydroids or among Laminaria.

<u>Classification</u>: Class Anopla, Order Palaeonemertea, Family Cephalothricidae

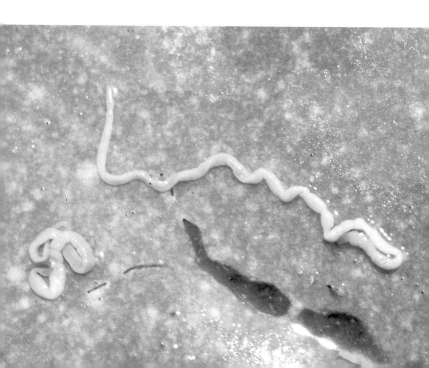

Phylum Annelida (segmented worms)

Annelids are segmented worms. The body is divided into many parts. The nervous, circulatory and excretory systems are also segmented. The gut runs through the body from the mouth to the anus. Most annelids respire through the skin. A few have gills. Segmented worms can be found in most habitats. This phylum includes the oligochaetes (the earthworms and freshwater worms), the hirudins (the leeches), and the polychaetes (most of the marine worms).

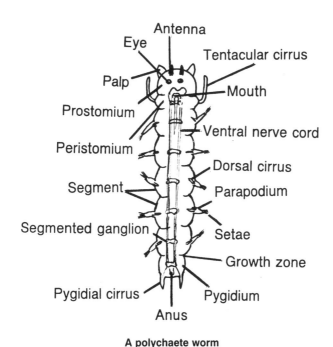

A polychaete worm

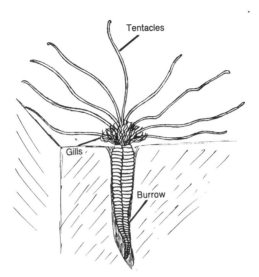

A sedentary polychaete worm

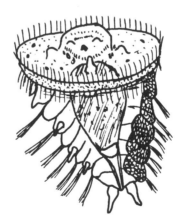

Trochophore larvae of an annelid

Class Polychaetes (fireworms, tubeworms, lugworms)

These worms form a large group of marine animals. Segmentation is clearly defined in most polychaetes. They possess bristles which are usually grouped on paired appendages found on each segment. Polychaetes can either be free moving or sessile, however this definition is not always clear. The free moving polychaetes can be pelagic, soft substrate burrowers, or crawlers. They usually move by body contractions and/or the use of parapodia. Their head possess sensory appendages. Sedentary species usually construct burrows or tubes from which they project their head and/or feeding apparatus. They may have gills or feeding tentacles; however, they lack sensory appendages. Some species construct complex burrows. Polychaetes utilize a variety of food types. They can be predators, scavengers, grazers, filter feeders or deposit feeders, to name a few. Although some polychaetes reproduce asexually, the majority of them reproduce sexually. Usually the sexes are separate, fertilization is mostly external and the eggs of many species are planktonic. Some species brood their eggs in their tubes, whereas others brood them in their operculum. Eggs develop into trochophore larva that metamorphosize into juveniles. Many epitokous polychaetes swim to the surface and release their gametes synchronously. This event, called swarming can be visually impressive. Some species bioluminesce during nighttime swarming creating a spectacular display.

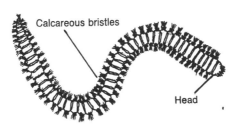

The fireworm, *Hermodice caruncalata*

Hermodice carunculata

Common names: fireworm (En), skoulipetra (Gr), feuerwurm (G)

Description/Biology: Fireworms have a clearly segmented body and are slightly flattened. Along each side of the body there are groups of white calcareous bristles that are hollow and contain poison. When threatened, the bristles become more prominent (upper photo). These worms are mostly active at night feeding on dead organisms and preying upon corals and anemones. They are found on rocky bottoms between 1 m and 50 m depth. Fireworms can reach 30 cm in length. This species can also be found in other tropical and temperate waters outside of the Mediterranean.

Classification: Order Amphinomida, Family Amphinomidae

Comments: Fireworms should not be handled without gloves. Contact with the bristles causes irritation.

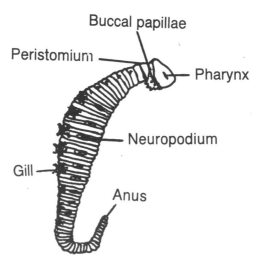

Buccal papillae

Peristomium

Pharynx

Neuropodium

Gill

Anus

The blow lug, *Arenicola*

Arenicola marina

Common names: blow lug (En), skouliki (Gr), wurm (G)

Description/Biology: When young, blow lugs are redish-pink. The adults have a greenish-yellow colour. If handled they stain yellowish. The tail region is narrower and more fragile than the rest of the body and it does not posses any chaetae or gills. The body is cylindrical and made up of short segments. These polychaetes grow to be 20 cm in length and live in burrows in clean to muddy sand on the lower shore.

Classification: Order Capitellida, Family Arenicolidae

Comments: This is an important fish bait species.

Eupolymnia nebulosa

Common names: misty tubeworm (En), skouliki (Gr), nebel-
eupolymnie (G)

Description/Biology: These worms live in a buried mucous tube
which has sand and other particles attached to it. Usually the
most visible part of the worm are the long (up to 30 cm)
tentacle-like white filaments. These usually radiate from the
tube and are used to trap small organic particles on the sticky
surfaces. The filaments contract to deliver captured food to
the mouth or when the worm is disturbed. Misty tubeworms
are diurnal and live on hard substrates from 1 m to 15 m. They
can grow up to 15 cm in length. This species can also be found
in other tropical and temperate waters outside of the
Mediterranean.

Classification: Order Terebellida, Family Terebellidae

Spirographis spallanzani

Common names: feather dusterworm (En), skouliki (Gr),
schraubensabelle (G)

Description/Biology: Feather dusterworms build non-calcareous soft
tubes. From the top of the tube extend radioles in a single
whorl. The radioles are used to trap fine organic particles
from the water column, which are transported to the mouth
by cilia. The radioles can quickly be retracted when the worm
senses danger. Feather dusterworms are found on or around
hard substrates from 2 m and 50 m. They are diurnal and the
tubes can grow up to 30 cm in length.

Classification: Order Sabellida, Family Sabellidae

Protula intestinum

Common names: red calcareous tubeworm (En), skouliki (Gr), roter
kalkrohrenwurm (G)

Description/Biology: This species produces a white calcareous tube
that can become covered by algae. Calcium secreted by two
glands is mixed with organic matter secreted from the ventral
surface. This matrix forms the structure of the tubes.
Tubeworms are filter feeders, using the same feeding
technique as *Spirographis spallanzani*. They also have the
ability to rapidly withdraw into their shells upon sensing
danger. Calcareous tubeworms are diurnal and can be found
on hard substrates at depths between 1 m and 50 m. The
radioles can reach a length of 3 cm.

Classification: Order Sabellida, Family Serpulidae

Phylum Echiura (spoonworms)

These marine organisms are cylindrically shaped and have a body that is made up of a trunk and a proboscis. They usually inhabit burrows in soft substrates or crevices on hard bottoms. The ventrally ciliated proboscis is extended for feeding. Most echiurans are deposit feeders. Food that drops onto the proboscis is transported by the cilia to the mouth. Echiurans reproduce sexually. The sexes are separate and sexual dimorphism exists in some species. Gametes are usually shed into the water column. The fertilized eggs develop into a planktonic trochophore larva, that eventually settles and metamorphosizes into a juvenile echiuran. Most species live in shallow waters.

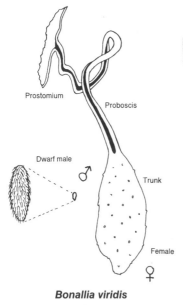

Prostomium

Proboscis

Dwarf male ♂

Trunk

Female ♀

Bonallia viridis

Bonellia viridis

Common names: green bonellia (En), skouliki (Gr), russel-oder
   igelwum (G)

Description/Biology: Green bonellia are found in crevices. They
   usually extend their long green proboscis over the
   surrounding substrate at night. The main trunk, that is usually
   hidden, can reach a size of 8 cm in length, whereas the
   proboscis can be as long as 2 m. The terminal portion of the
   proboscis is split into two parts. When extended, the
   proboscis is curled ventrally so that the cilia are able to
   channel food particles to the mouth. This species exhibits
   distinct sexual dimorphism. The males (about 3 mm) are
   much smaller than the females and live attached to the genital
   sack of the female. Males do not posses a mouth, anus or
   vascular system. They are nourished by the surrounding
   female fluids. Males are essentially a sack of sperm, they
   fertilize the eggs and secrete a material that glues the eggs
   together. Fertilized eggs develop into trochophore larvae that
   eventually settle. Larvae that settle on a female proboscis
   metamorphose into males and attach to the proboscis.
   Eventually they migrate to the genital sacks of the female.
   Larvae that don't settle on a proboscis become females.

Classification: Order Bonelliida, Family Bonelliidae

Comments: The green colour of *Bonellia viridis* is due to a pigment
   (bonellin) that is thought to have antibiotic properties.

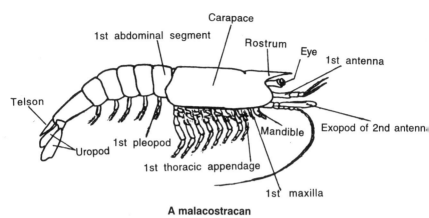

A malacostracan

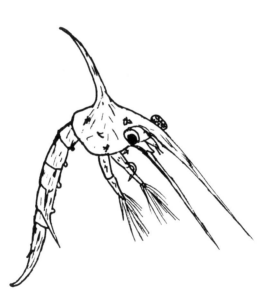

Decapod larvae, zoeal stage

Phylum Arthropoda (arthropods)

This is the largest phylum of animals and contains more species than all the other phyla combined. The main characteristics of this phylum are jointed legs and a segmented body covered by a chitinous exoskeleton. The bodies of most arthropods are divided into three sections: the head (cephalum), chest (thorax) and abdomen. Often the head and chest may be fused together to form the cephalothorax. In order to increase in size, arthropods grow a soft expandable exoskeleton and crawl out of the old hard shell. The soft exoskeleton hardens to a larger size shell that accommodates the growing arthropod. During this process (molting) an arthropod without its hard protective cover is vulnerable and will therefore hide until the new exoskeleton is hardened. Most arthropods are terrestrial, these include all the insects, spiders, mites and centipedes. Marine representatives include the barnacles, isopods, shrimps, lobsters and crabs.

Class Crustacea (barnacles, isopods, shrimps, lobsters, crabs)

Most crustaceans are aquatic organisms and most marine arthropods are crustaceans. This group has a wide range of shapes and sizes ranging from small parasitic isopods to large 1 m long lobsters and spider crabs with legs over 1.5 m long. They posses two pairs of antennae and compound eyes. Respiration is achieved either by gills or by diffusion through their body surface. Usually the sexes are distinct and the young pass through at least one larval stage. Many crustaceans are important commercially and are considered delicacies. For this reason special fishing methods have been developed to target them. Because of their economical importance, in the last few years there has been a rapid development in crustacean aquaculture.

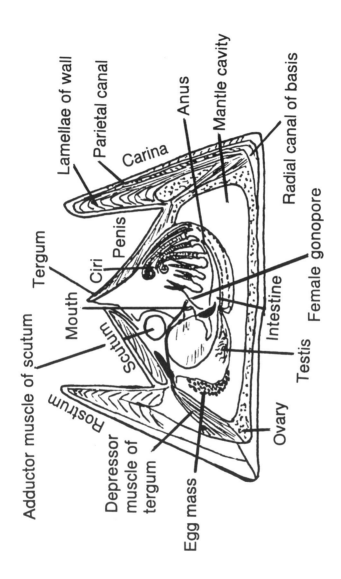

A vertical section of a barnacle

Lamellae of wall
Parietal canal
Carina
Anus
Mantle cavity
Radial canal of basis
Penis
Tergum
Ciri
Mouth
Scutum
Rostrum
Adductor muscle of scutum
Depressor muscle of tergum
Egg mass
Ovary
Testis
Female gonopore
Intestine

Chthamalus stellatus

Common names: barnacle (En), bibiki (Gr), seepoke (G)

Description/Biology: Even though they have a free swimming larval stage these crustaceans settle on hard substrates as adults. They attach themselves to a firm object with the help of a cement like substance secreted by a gland found on the base of their antennae. After attachment they secrete a calcareous nearly conical shell with six overlapping plates. The pair of plates that close the opening of this protective structure can be opened and closed at the midline allowing 6 pairs of cirri to emerge. The cirri look like miniature feathers and trap microorganisms which are then passed to the mouth of the barnacle found within the protective shell. Because barnacles live primarily in the intertidal zone and in areas frequently exposed to air, their tightly closed shells offers protection not only against predators but also against dry conditions encountered during low tides. Barnacles are hermaphrodites and can be found in fairly large colonies. This species reaches sizes of about 1 cm in diameter.

Classification: Subclass Maxillopoda, Order Thoracica, Family Chthamalidae

Comments: Barnacles are considered a nuisance to boat owners because by colonizing boat hulls they increase the hydro-dynamic resistance of the hull and thus decrease the performance of the boat. A lot of money has been spent developing special toxic paints to protect boat hulls from barnacle colonization.

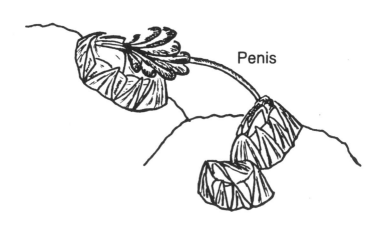

Penis

Barnacles copulating

Anilocra physodes

<u>Common names:</u> fish louse (En), psira (Gr), fischassel (G)

<u>Description/Biology:</u> The fish louse is an ectoparasite that attaches to fishes. It possesses a dorsal ventrally flattened body with a small shield shaped head. It has 7 thoracic and 5 abdominal segments. It uses its six pairs of legs which end in hook like appendages to grab and hold onto its host. The dorsal side is dark grey to black and the ventral side is lighter. Although the fish louse is most frequently observed attached to fishes, it does have the ability to swim. Upon attaching to a host, the mandibles pierce and damage the skin.

<u>Classification:</u> Order Isopoda, Family Cymothoidae

<u>Comments:</u> Although parasitic isopods do not usually kill their host, they may make their host an easier target to predators, diseases and other parasites.

Gammarus sp.

<u>Common names:</u> seaweed hoppers (En), psilos tis thallasas (Gr)

<u>Description/Biology:</u> These small organisms are commonly found under rocks and seaweeds along beaches and the intertidal zone. Most species are less than 1.5 cm long and even under magnification are difficult to tell apart. Most species have the characteristic ability to jump, hence their common name. Most feed on detritus.

<u>Classification:</u> Subclass Malacostraca, Order Amphipoda, Family Gammaridae

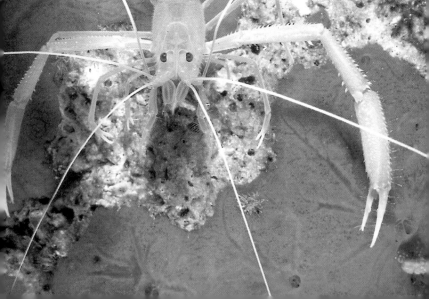

Stenopus spinosus

<u>Common names:</u> boxer shrimp (En), garida (Gr), scherengarnele (G)

<u>Description/Biology:</u> The boxer shrimp belongs to the same family as the cleaner shrimp that are found in tropical waters. Unlike its relatives, the boxer shrimp has not been documented to clean other organisms. The boxer shrimp has 3 pairs of characteristic long white antennae. As its order suggests, this arthropod has 10 legs (5 pairs). It is light orange-yellow in colour. These arthropods inhabit cracks and crevices from 1 m to 75 m depth and are most active at night. They usually grow to be 7 cm long. Boxer shrimp are scavengers.

<u>Classification:</u> Subclass Malacostraca, Order Decapoda, Family Stenopodidae

<u>Comments:</u> These creatures are rarely seen during the day. A diver at night will usually see the long antennae protruding from the hideout of the boxer shrimp. These shrimp are probably edible, but because of their low abundance they are not fished. Due to their striking colour and intricate bodies, shrimps belonging to the family Stenopodidae are valued by aquarists.

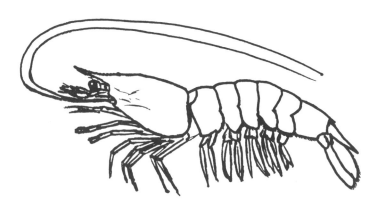

A shrimp

Palaemon elegans

Common names: common/rock pool prawn (En), garidaki (Gr), felsengarnele/steingarnele (G)

Description/Biology: These small prawns have a transparent body making them very elusive. Usually they are found in fairly large numbers in shallow waters and tide pools. Their body and appendages are faintly marked with black and yellow stripes. Like other decapods, female prawns with the help of pleopods will carry fertilized eggs underneath their abdomen. These arthropods are scavengers and cleaners. Sometimes a standing bather will notice these creatures nibbling at his or her feet. Common prawns are found from shallow tide pools to 2 m depth. These small decapods are rarely larger than 5 cm.

Classification: Subclass Malacostraca, Order Decapoda, Family Palaemonidae

Comments: Even though the common prawns are not as colourful as *Stenopus spp.*, they are very robust and thus suitable for aquariums and scientific research. Their small size does not inhibit these delicious sweet tasting creatures from becoming ouzo appetizers in some Greek areas.

Homarus gammarus

Common names: lobster (En), astakocaravida (Gr), hummer (G)

Description/Biology: This species is very similar to the American lobster (*Homarus americanus*). It has two large unequal size claws. The larger of which has blunt teeth and is used for crushing hard shelled prey. The other claw is thinner and narrower with sharper teeth and is used for catching and slicing softer and more elusive prey. These decapods are greenish brown and turn red when boiled. Lobsters are benthic. During the day they hide in holes and crevices. At night lobsters hunt and scavenge for food. They can be found in rocky areas at depths between 2 m and 50 m (see neighboring photo). Larger individuals live in deeper waters. The usual catch size is between 30 cm and 50 cm; however, lobsters have been recorded to reach sizes greater than 1 m. Although lobsters usually move slowly using their legs, they have the ability to move fast over short distances with the help of their powerful tail. Dominant males occupy the "best" holes. Females are attracted to the holes by the pheromone in the males urine. Once inside a male's hole, the female will molt and mate. Once her new shell is hardened, she moves out of the hole. Dominant females will visit first the dominant male.

Classification: Subclass Malacostraca, Order Decapoda, Family Nephropsidae

Comments: H. gammarus is not very common in the Mediterranean. Occasionally it can be found in Greek seafood markets where it is highly prized for its delicious flesh. Usually lobster is boiled and can be prepared in various different ways. A glass of white wine or retsina is always a good companion to a lobster dish. Lobsters should be handled by the carapace. The powerful snap of the lobsters tail can seriously injure a hand.

Palinurus elephas

Common names: spiny lobster (En), astakos (Gr), languste (G)

Description/Biology: The spiny lobster is much more common in the Eastern Mediterranean than *Homarus gammarus. P. elephas* does not have two large claws, however two of the four antennae are long with a thick spiny base. These antennae can be used to distract predators while the lobster backs off into a crevice. It has a reddish - brown colour and stockier body than *H. gammarus*. A nocturnal species, the spiny lobster leaves its crevices at night to scavenge on the sea floor. It inhabits rocky areas at depths from 10 m down to 80 m. The lifestyle of spiny lobster is more or less similar to that of other decapods. It attains sizes larger than 80 cm but most of the captured spiny lobsters are between 30 cm to 40 cm.

Classification: Subclass Malacostraca, Order Decapoda, Family Palinuridae

Comments: Spiny lobster can be found in Greek seafood markets where it is highly prized for its delicious flesh. Usually spiny lobster is boiled, however it can be prepared in various ways. A glass of white wine or retsina is always a good companion to a lobster dish. Spiny lobsters should be handled by the carapace. The powerful snap of the spiny lobsters tail can be a "weapon" and seriously injure a hand. Injuries from a spiny lobster's tail are much more serious than the ones inflicted by the tail of Homarus gammarus.

Scyllarides sp.

<u>Common names:</u> Spanish lobster (En), karakoukos (Gr), grober barenkrebs (G)

<u>Description/Biology:</u> Spanish lobsters are smaller than the other two described species. They do not possess any "weapons" except for their hard shells. Their two front antennae are flattened and form broad plates. They are more agile than the other two lobsters. Spanish lobsters are benthic scavengers and are most active at night while they hide in caves and crevices during day. They can reach sizes larger than 30 cm but the average size of individuals caught in nets is around 15 cm. They can be found on rocky substrates at depths between 10 m and about 60 m.

<u>Classification:</u> Subclass Malacostraca, Order Decapoda, Family Scyllaridae

<u>Comments:</u> Although these lobsters are edible, they are not found in Greek seafood markets.

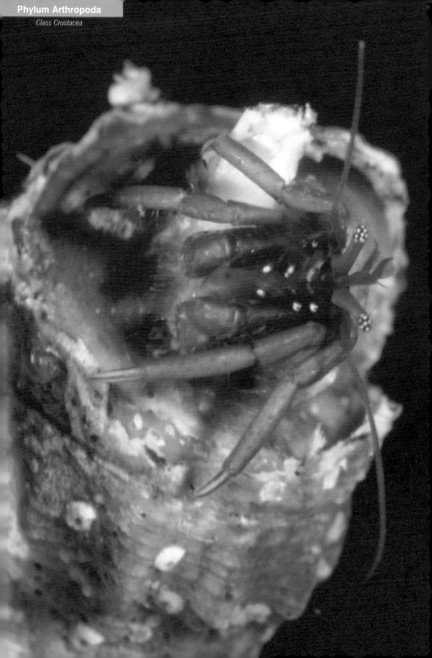

Calcinus sp.

<u>Common names:</u> hermit crab (En), carkinari (Gr), einsiedler (G)

<u>Description/Biology:</u> Hermit crabs have a soft asymmetrical abdomen. Partly for this reason, they usually occupy empty gastropod shells for protection. As they grow in size they abandon their old shells and move into a larger shell. Their carapace is oval and the first pair of legs are formed into chelipeds. The third, fourth and fifth pairs of legs are not chelate. Hermit crabs are benthic and can be found on most substrates. They are scavengers and are most active at night. This small hermit crab grows to about 2 cm (body length) and lives in shallow waters down to 5 m. It is active during the day as well as at night.

<u>Classification:</u> Subclass Malacostraca, Order Decapoda, Family Paguridae

<u>Comments:</u> Hermit crabs make good fishing bait.

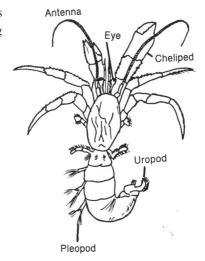

Dorsal view of a hermit crab out of its shell

Dardanus calidus

<u>Common names:</u> red hermit crab (En), carkinari (Gr), einsiedler-
krebse (G)

<u>Description/Biology:</u> This hermit crab usually has symbiotic
relationships with sea anemones. The sea anemone offers the
crab protection and the crab transports the sea anemone to
areas of greater food abundance. In the picture of this species,
as the hermit crab was being photographed, the sea anemone
released a sticky purple thread for protection. The red hermit
crab grows to 8 cm (body length) and can be found from
shallow waters to around 50 m depth.

<u>Classification:</u> Subclass Malacostraca, Order Decapoda, Family
Paguridae

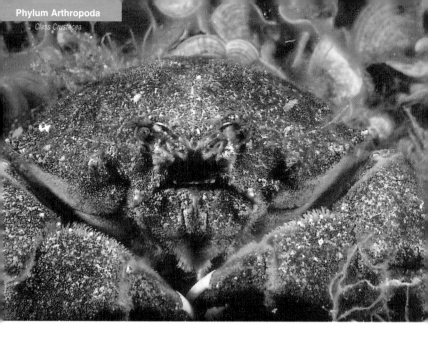

Phylum Arthropoda
Class Crustacea

Dromia personata

<u>Common names:</u> sponge crab (En), sfougarocavouras (Gr), krabbe (G)

<u>Description/Biology:</u> Crabs have a symmetrical abdomen that is reduced and tightly flexed beneath the cephalothorax. Their first pair of legs form strong chelipeds, their third, fourth and fifth pairs of legs are not chelate. The carapace is usually more broad than long. The sponge crab is dark brown with pink chelae. The body is covered with short velvet like hairs. The chelae are stout and equal in size. Females have smaller chelae than males. The body is often covered in a sponge providing camouflage (hence the name). The carapace breadth of males can reach 10 cm. They are found in rocky environments and in caves from 1m to 100 m depth. Sponge crabs are predators and scavengers and are more active at night.

<u>Classification:</u> Superorder Eucarida, Order Decapoda, Family Dromiidae

<u>Comments:</u> The sponge crab can be very elusive and it can blend into its surrounding environment, using sponges for camouflage.

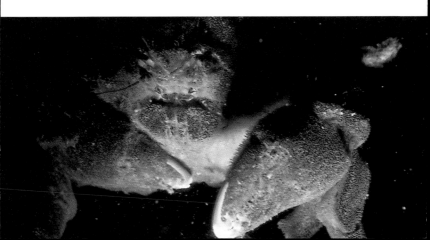

Munida rugosa

<u>Common names:</u> squat lobster (En), astakaki (Gr), furchenkrebs (G)

<u>Description/Biology:</u> The squat lobster is a close relative of the hermit crabs. The abdomen is curled underneath the cephalothorax. The first pair of legs are formed into very long chelipeds. There is a pair of long antennae. The tail fan is well developed. They are pinkish red with redish edges and groves. The carapace has some spines that are redish in colour. These decapods usually inhabit rocky areas with small crevices. Squat lobsters are scavengers, active mostly at night. They grow to around 7 cm in body length and are usually found at depths deeper than 10 m down to 150 m on stony bottoms.

<u>Classification:</u> Subclass Malacostraca, Order Decapoda, Family Galatheidae

<u>Comments:</u> Squat lobsters are edible. However,they are not easy to find and are not sold at seafood markets.

Herbstia condyliata

<u>Common names:</u> decorative crab (En), kavouras (Gr), krabbe (G)

<u>Description/Biology:</u> Decorative crabs have velcro like hairs on their carapace and legs that are used to secure algae and other live or dead materials. The "decorations" serve as effective camouflage (as can be seen in the picture).

These crabs are fairly small, growing to around 5 cm (carapace length). They are usually found in caves and crevices in waters deeper than 10 m.

<u>Classification:</u> Subclass Malacostraca, Order Decapoda, Family Majidae

<u>Comments:</u> These small crabs have excellent camouflage and often take on interesting appearances as they decorate themselves with a variety of items.

Maja squinado

Common names: spider crab (En), atchiganos (Gr), seespinne (G)

Description/Biology: Spider crabs have a more or less triangular carapace and long legs. They are active during the day and night and are usually found on soft substrate at depths from 1 m to 50 m. Spider crabs of the genus *Maja* can grow to around 25 cm (carapace length). This spider crab is brownish red or yellowish in colour and has long legs and chelipeds. The chelipeds can reach 45 cm in length and in males, are thicker than the legs. The body is often covered in attached algae.

Classification: Subclass Malacostraca, Order Decapoda, Family Majidae

Comments: Spider crabs are often pulled up in fishermen's nets as they scavenge for the captured fishes. Spider crabs are edible. A relative of the spider crab is the king crab that is considered a delicacy.

Liocarcinus corrugatus

Common names: wrinkled swimming crab (En), kavouras (Gr), krabbe (G)

Description/Biology: The last pair of legs are more or less flattened and paddle like. This pair of legs is used mostly for swimming. The area on the carapace between the eyes has three similarly sized teeth. The carapace and parts of the limbs are covered by transverse rows of hairs arising from ridges. The carapace can reaches approximately 4.5 cm in length and 4 cm in width. The colour is brownish red with patches of red or yellow. The wrinkled swimming crab is found on hard substrates from 1 m to 100 m depth. Like most other crabs, it is a predator and scavenger and is most active at night.

Classification: Subclass Malacostraca, Order Decapoda, Family Portunidae

Comments: Although they are edible, wrinkled swimming crabs are not very common and are not found in Greek markets. Small individuals can be used as bait.

Carcinus aestuarii

Common names: shore crab (En), kavouras (Gr), mittelmeer strandkrabbe (G)

Description/Biology: The last pair of legs are modified into paddle shapes used for swimming. The carapace is fairly round and is notched around the edges. This crab is olive-brownish on the dorsal surface and whitish on the ventral surface. It is usually found in tide pools and shallow waters down to 2 m. Crabs are common along the coastline. They are active both day and night. The carapace can reach 5 cm in width.

Classification: Subclass Malacostraca, Order Decapoda, Family Portunidae

Comments: This crab can be used as bait.

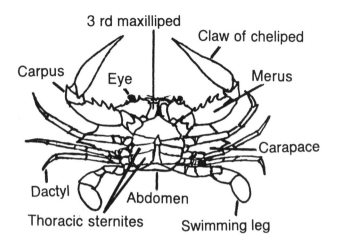

Ventral view of a brachyuran crab, Family Portunidae

Carcinus maenas

Common names: green crab (En), kavouras (Gr), krabbe (G)

Description/Biology: Green crabs are common along the coastline. They are one of the species that may leave the water to scavenge on the seashore. The dorsal surface of their body is a dark green and the ventral side of males and juveniles is yellow to cream; in females it is red-orange . The last pair of legs are slightly flattened. The carapace is rounded with 5 marginal and 3 frontal teeth between the eyes. Green crabs are usually intertidal. They are active both during the day and night. They may grow up to 8 cm in carapace width. Green crabs are usually found from very shallow waters down to 3 m. Although the green crab is a European species it has been introduced to many other areas such as America, Hawaii and Australia.

Classification: Subclass Malacostraca, Order Decapoda, Family Portunidae

Comments: Although they are edible and common, green crabs are not usually found in Greek seafood markets. Small individuals can be used as bait.

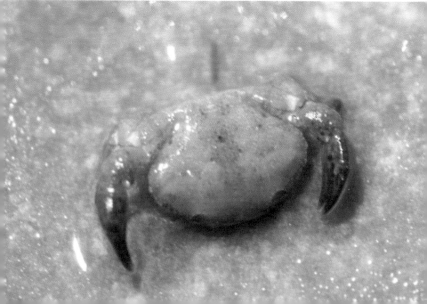

Eriphia verrucosa

<u>Common names:</u> warty crab (En), pagouras (Gr), italienischer taschenkrebs (G)

<u>Description/Biology:</u> This species has a thick exoskeleton and very powerful claws. As the name suggests, the front part of the carapace and chelipeds possess short thick spines. The legs, chelipeds and outer edges of the carapace are covered with thick hairs. The warty crab is brown to dark brown in colour. It lives in crevices in shallow waters down to 10 m. It is active at night and grows to 10 cm in carapace width.

<u>Classification:</u> Subclass Malacostraca, Order Decapoda, Family Xanthidae

<u>Comments:</u> Although they are very tasty and fairly common, warty crabs are not usually found in Greek markets. Small individuals can be used as bait.

Xantho poressa

<u>Common names:</u> stone crab (En), kavouraki (G), steinkrabbe (G)

<u>Description/Biology:</u> This small crab has an oval carapace and relatively large, powerful claws. The dorsal side of the stone crab is sandy brown with black spots. The claws are black. It is active at night and during the day it usually hides in rocky and sandy environments. The stone crab can be found in shallow waters down to 10 m. It grows to 3 cm in carapace width.

<u>Classification:</u> Subclass Malacostraca, Order Decapoda, Family Xanthidae

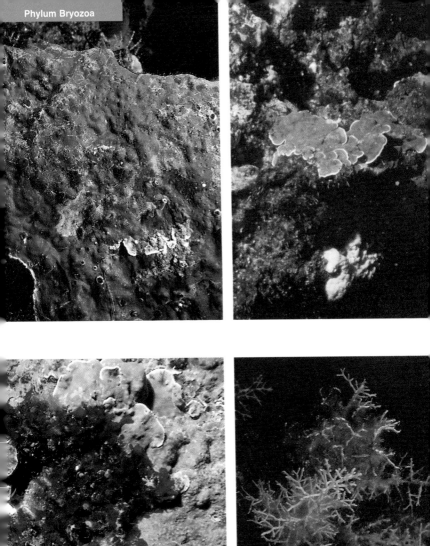

Phylum Bryozoa (bryozoans)

Bryozoans are sessile colony forming animals. This phylum consists of approximately 5000 species; however, because bryozoans are so small, they usually go unnoticed. Colonies are made up of zoids that are approximately 0.5 mm in length. Some species form encrusting plates, while others form small delicate fans. Many species are calcified and sometimes their colonies appear similar to small coral colonies. Zoids are often polymorphic and colonies are hermaphrodite. Eggs are usually brood, either in the coelom or externally. Zoids can be box-like, egg shaped, or cylindrical. They are comprised of a trunk and a tentacle sheath that bears the lophophore. The lophophore has a series of slender ciliated tentacles and is used for filter feeding phytoplankton from the water column.

Bryozoans are found on many different types of hard surfaces such as rocks, coral, wood, shells and sand grains. They can also be found on other surfaces such as seaweeds and eelgrasses. Like most filter feeding organisms they thrive in environments with high amounts of water movement, such as channels and straits. They are considered an important group for fouling ships' hulls.

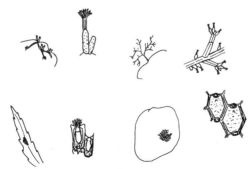

Several types of marine bryozoans

Sertella beaniana

<u>Common names:</u> mermaid's veil (En), corali (Gr), neptunschleier (G)

<u>Description/Biology:</u> This bryozoan forms an orange-yellow intricate lace like structure. The colony is formed from a single zoid that attaches itself to the substrate. The colony develops from asexual germination. The diameter of the "veil" is around 5 cm and can be found on hard substrates from 5 m to 50 m depth.

<u>Classification:</u> Order Cheilostomata, Family Reteporidae

<u>Comments:</u> The mermaid's vail is sometimes mistaken for coral.

Phylum Echinodermata (echinoderms)

The echinoderms are only found in marine habitats. Most of them are benthic. Their body has pentamerous radial symmetry that is derived from bilateral symmetry. They have an internal calcareous skeleton which typically has projecting spines or tubercles. The phylum name (spiny skin) describes the surface of echinoderms. In some echinoderms, such as sea urchins and sand dollars, the skeleton is continuous and in others, such as starfish, it forms articulated plates. Echinoderms have a water-vascular system that moves water through a system of body canals connected to the external environment through the madreporite. This system enables echinoderms to move around as well as to obtain food. Sexes are usually separate and fertilization is external. The larvae spend part of their lives as planktonic organisms. Several members of this group (e.g. starfish) have the ability to regenerate parts of their body.

Class Asteroidea (starfish/sea stars)

Starfish are mainly carnivorous, benthic echinoderms that usually have 5 stiff arms radiating from a central disk. The arms are usually narrow at the tips and increase in width towards the central disk. A typical starfish body is relatively flat with distinct ventral and dorsal sides. The mouth is located on the ventral side in the center of the disk. Grooves extend from the mouth to the tip of each arm. These ambulacral grooves contain 2 or 4 rows of tubular feet that are used for movement, feeding and sensory reception. The stomach fills most of the interior of the central disk and in many species it can be pushed out through the mouth to digest food outside of the body. This feeding mode is especially useful for feeding on bivalves. In such cases the bivalve is held open by the tube feet and the stomach is inserted into the bivalve through the narrow opening. The stomach excretes gastric juices that digest the prey's soft body. Due to their feeding habits, starfish are considered a pest by some bivalve culturists. Some starfish are very colourful. They are not considered to be edible.

Astropecten aranciacus

Common names: red comb starfish (En), asterias (Gr), roter kamm-
seestern (G)

Description/Biology: Along the flattened arms of this starfish are
pronounced white spines. These arms narrow towards the
ends. The perimeter of the starfish is made up of hard grayish
plate like structures. The dorsal surface is abrasive due to
numerous small spines. The upper side of the starfish is
brownish-red and the underside is whitish-yellow. It is active
both during the day and during the night. It can be found on
sandy and muddy bottoms from 5 m to 30 m depth. It grows
to be 35 cm in diameter.

Classification: Order Phanerozonia, Family Astropectinidae

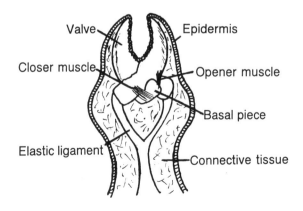

Scissor-type pedicelaria from a starfish, *Asterias*

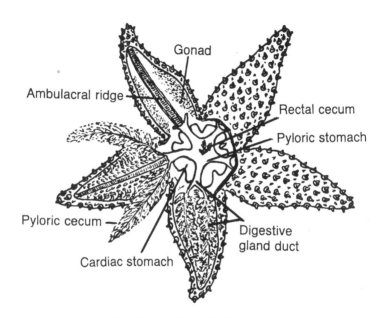

Gonad

Ambulacral ridge

Rectal cecum

Pyloric stomach

Pyloric cecum

Digestive gland duct

Cardiac stomach

**Aboral view of a starfish, *Asterias*,
with the arms at various stages of dissection**

Astropecten bispinosus

<u>Common names:</u> brown comb starfish (En), asterias (Gr), brauner kamm-seestern (G)

<u>Description/Biology:</u> The brown comb starfish is smaller than the red comb starfish. Along the arms are pronounced brownish-ivory spines. The flattened arms narrow towards the ends and are darker along the center. The upper side of the starfish is brownish-grey and the underside is whitish-yellow. Active both during the day and during the night. It can be found on sandy and muddy bottoms from 5 m to 30 m depth. To move on these types of substrates, it uses specialized tube feet with pointed tips. Other starfishes that live on hard substrates have tube feet with suction pads. It grows to 15 cm in diameter.

Classification: Order Phanerozonia, Family Astropectinidae

Luidia ciliaris

Common names: seven rayed starfish (En), eptaktinos asterias (Gr),
siebenarmiger grobplatten-seestern (G)

Description/Biology: This starfish can be easily identified because it
has seven arms. The slightly flattened arms maintain the same
width for most of their length, becoming slender close to the
tips. Along each arm are visible spines. The arms are very
fragile and will break off easily when the starfish is handled;
however, they will regenerate quickly. The dorsal and ventral
sides are orange-red. This starfish is nocturnal and is found on
soft bottoms at depths between 5 m and 100 m. It can reach a
diameter of 40 cm. This starfish looks very similar to *Luidia
sarsi* which only has five arms.

Classification: Order Phanerozonia, Family Luidiidae

Echinaster seposistus

<u>Common names:</u> red starfish (En), kokinos asterias (Gr), purpustern (G)

<u>Description/Biology:</u> The red starfish has 5 cylindrical arms and a rough surface. The central disk is relatively small. Its red colour makes it easy to recognize. It can be found on most benthic habitats from 2 to 50 m depth. It grows to around 25 cm in diameter. The red starfish is carnivorous, feeding mostly on benthic bivalves and gastropods.

<u>Classification:</u> Order Spinulosida, Family Echinasteridae

Marthasterias glacialis

<u>Common names:</u> spiny starfish (En), akanthasterias (Gr), eisseestem (G)

<u>Description/Biology:</u> As the name indicates, this starfish has a very spiny body. The dorsal surface also possess distinct pedicellaria. The arms are fairly cylindrical. Its colour is a mixture of brown, orange and olive-green. During the day, it usually hides under rocks. It is an active scavenger and predator at night. This starfish is known to prey on sea urchins as well as bivalves. It is predominantly found on most types of hard substrates between 2 m and 40 m. It grows to reach a diameter of 40 cm.

<u>Classification:</u> Order Forcipulata, Family Asteriidae

Sphaeriodiscus placenta

Common names: cushion starfish (En), asterias (Gr), fladenstern (G)

Description/Biology: The 5 rather flattened arms are connected for most of their length. There is a ridge of thickened tissue around the edge of the body. Its colour ranges from yellow to orange-red. The cushion starfish usually reaches a diameter of around 10 cm. It lives on rocky substrates at depths between 20 and 100 m.

Classification: Order Valvatida, Family Goniasteridae

Phylum Echinodermata
Class Ophiuroidea

Class Ophiuroidea (brittle stars)

This is the most speciouse group of echinoderms. They are generally smaller than asteroids and have clearly distinguishable arms. The central disk is usually round and flattened. There is no ambularcal groove and the tubular feet are not used for locomotion. They usually have 5 long cylindrical arms (2 to 20 times the disk diameter). Brittle stars are the fastest echinoderms. The arms are more flexible than those of starfishes allowing them to hide in small crevices. The skeleton is covered with spines. Sometimes these spines are small and inconspicuous. The arms can detach easily if grabbed by a predator. Ophiuroids are mostly nocturnal scavengers and predators. They are found on most substrates from shallow waters down to abyssal depths. Some species filter sea water for planktonic organisms.

Ophioderma longicaudum

Common names: brown brittle-star (En), ofiouros (Gr), brauner
    schlangenstern (G)

Description/Biology: This brittle-star is coloured different shades of
    brown and red. In the daytime it hides under rocks and in
    crevices. Occasionally one or two arms are left protruding
    from the hiding place. When disturbed, the brittle-stars will
    move quickly and find new cover. These are nocturnal
    predators and scavengers. They are usually found in rocky
    areas at depths from 1 m to 50 m. This is one of the largest
    brittle stars reaching a size of 25 cm in diameter.

Classification: Order Ophiurae, Family Ophiodermatidae

Ophiothrix quinquemaculata

<u>Common names:</u> common brittle-star (En), ofiouros (Gr), stacheliger schangenstern (G)

<u>Description/Biology:</u> The common brittle-star has numerous long spines that offer protection from predators. It is found mostly on structures on muddy bottoms and occasionally it is found on rocky bottoms. This brittle star traps passing food with its arms. It is active both during the day and during the night. The common brittle-star can reach 12 cm in diameter and can be found from 2 m to 100 m depth.

<u>Classification:</u> Order Ophiurae, Family Ophiotrichidae

Class Echinoidea (sea urchins, sand dollars)

Echinoids have a spherical or disk shaped body without arms. The rigid calcareous shell or test is covered with tubercles which are the base for movable spines. In sea urchins, the spines are usually long and heavy, whereas in sand dollars the spines are numerous and tiny giving them a furry look. The test is divided into five ambulacral segments which start at the anus, terminate at the mouth, and are separated by five interambulacral areas. In addition to spines, the ambulacral segments contain tube feet which are connected to the internal water vascular systems by canals. These canals can be seen on the test of a dead sea urchin as small paired perforations. Echinoids also have pedicellaria which are composed of a long stalk ending in a three part jaw like structure. Some types of pedicellaria contain poisonous glands. These are usually found in tropical and subtropical species. Urchins usually posses more than one type of pedicellaria which they use for defense or cleaning the body surface. Movement of the pedicellaria and spines is attained by the use of muscles. Echinoids posses a remarkably complex mechanical scraping apparatus. Known as Aristotle's lantern, the structure is composed of 5 spear shaped magnesium enriched calcareous plates and numerous other smaller pieces. Special muscles protrude and retract the lantern through the mouth and close and open the five plates. Although the apparatus is used primarily for scraping algae, boring sea urchins can use it to dig cavities in rocks or rocky substrates. Echinoids can be found on or in a variety of substrates and are primarily grazers, scavengers and deposit feeders. Fertilization is external. Gametes are released into the water.

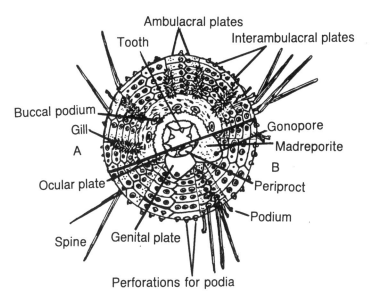

Ambulacral plates

Tooth

Interambulacral plates

Buccal podium

Gill

A

Gonopore

Madreporite

B

Ocular plate

Periproct

Podium

Spine

Genital plate

Perforations for podia

Oral (A) and aboral (B) view of a sea urchin

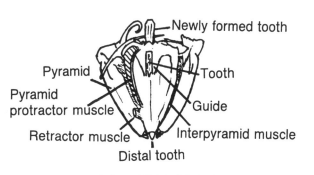

Newly formed tooth

Pyramid

Tooth

Pyramid protractor muscle

Guide

Retractor muscle

Interpyramid muscle

Distal tooth

Lateral view of Aristotle's lantern

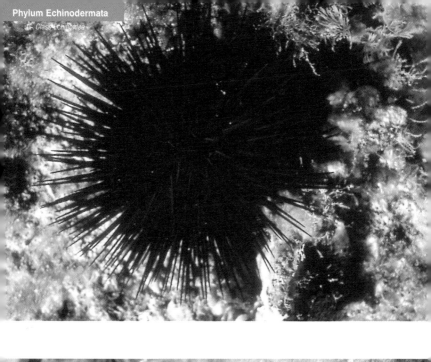

Arbacia lixula

Common names: black sea urchin (En), mavros ahinos (Gr), schwarzer seeigel (G)

Description/Biology: This sea urchin is black and morphologically resembles the red sea urchin. However, the two species can easily be distinguished underwater because the black sea urchin does not cover itself with objects, as does the red. This is believed to be because the black sea urchin lacks suckers on the tube feet on its upper side. It is active mostly at night, scraping algae off rocks as well as scavenging on a variety of benthic food items. The black sea urchin grows to 8 cm. It can be found on rocky substrates from shallow waters less than 1 m all the way down to 30 m depth.

Classification: Order Arbacioida, Family Arbaciidae

Comments: The black sea urchin may be edible. Sometimes it can be a nuisance to bathers. If stepped upon, the spines break after entering the body and can cause infections. To accelerate the removal of spines and reduce pain, we recommend applying olive oil to the injury. After a day or so, when the skin is soft, one can attempt to remove the spines with sterile tweezers and a pin.

Sphaerechinus granularis

Common names: violet sea urchin (En), agrioahinos (Gr), violetter
  seeigel (G)

Description/Biology: This sea urchin has shorter spines than the red
  and black sea urchin. It is purple in colour and the tips of the
  spines are white. This urchin has toxic pedicellaria that are
  used for defense. The toxin is not harmful to humans. The
  violet sea urchin is much larger than the other species
  described in this book. It can grow up to 12 cm in diameter.
  Its size, colour and short spines makes it easy to identify, even
  from a distance. It is active mostly at night feeding on algae
  and scavenging along the sea floor. Violet sea urchins can be
  found at depths from 4m to 70m.

Classification: Order Temnopleuroida, Family Toxopneustidae

Comments: Violet sea urchins are not edible. They are not a
  nuisance to bathers due to their depth distribution.

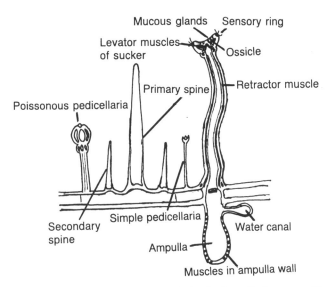

Schematic cross section through the body wall of asea urchin
showing the different types of spines and pedicellaria

Paracentrotus lividus

Common names: red sea urchin (En), kokinos ahinos (Gr),
steinseeigel (G)

Description/Biology: This sea urchin varies in colour from brown to
green to red. It usually deposits objects such as algae, stones
and shells on its body. Although the function of this behaviour
is unknown, it has been found to be related to photophobia.
The red sea urchin is active mostly at night scraping algae off
rocks as well as scavenging on a variety of benthic food items.
It grows to 8 cm in diameter. The red sea urchin can be found
on rocky substrates from shallow waters less than 1 m deep
down to 20 m.

Classification: Order Echinoida, Family Echinidae

Comments: Red sea urchins are edible and are usually eaten raw.
Sometimes they can be a nuisance to bathers. If stepped
upon, the spines break after entering the body and can cause
infections. To accelerate the removal of spines and reduce
pain, we recommend applying olive oil to the injury. After a
day or so, when the skin is soft, one can attempt to remove the
spines with sterile tweezers and a pin.

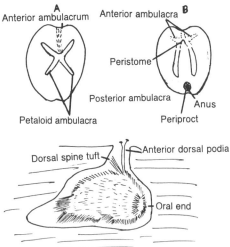

A
Anterior ambulacrum

B
Anterior ambulacra

Peristome

Posterior ambulacra

Anus

Petaloid ambulacra

Periproct

Dorsal spine tuft

Anterior dorsal podia

Oral end

**Heart urchins. A, aboral view. B, oral view.
The sea potato, *Echinocardium*, in a sand burrow.**

Echinocardium cordatum

<u>Common names:</u> sea potato (En), ahinos tis amou (Gr), herzseeigel (G)

<u>Description/Biology:</u> The sea potato is yellowish in colour and reaches sizes to about 9 cm in length. It is found buried 10 to 15 cm in sand, at depths of 1 m down to 230 m. This species is well dispersed and it can also be found in Japan, Australia and South Africa. It is active both during the day and night. The sea potato selectively feeds on organic particles in the sediment that are collected with modified podia on the oral surface.

<u>Classification:</u> Order Spatangoida, Family Spatangidae

<u>Comments:</u> Sea potatoes are probably not edible. Their test is more fragile than that of most other urchins. Sea potatoes often have symbiotic relationships with a bivalve mollusc, *Tellinya ferruginosa*, and an amphipod crustacean, *Urothoe marina*.

Spatangus purpureus

Common names: purple heart urchin (En), ahinos tis amou (Gr),
          herzseeigel (G)

Description/Biology: The purple heart urchin is usually buried in
          coarse sand or gravel from shallow waters down to 900 m. As
          the name implies, this urchin is heart shaped and deep violet
          in colour. Some of the long upper spines are white. Its empty
          test can be found on the sea floor. This urchin is bilaterally
          symmetrical and has numerous soft thin spines giving it a
          brushy appearance. The spines are used for locomotion and
          keep sand away from the body of the heart urchin. It is active
          both during the day and night. The purple heart urchin
          selectively feeds on organic particles in the sediment that are
          collected with modified podia on the oral surface. It reaches
          a length of 12 cm.

Classification: Order Spatangoida, Family Spatangidae

Comments: This heart urchin is probably not edible. Its test is more
          fragile than that of most other urchins. The bivalve mollusc,
          Mondacuta substriata, is often found attached to its spines.

Schizaster canaliferus

<u>Common names:</u> heart urchin (En), ahinos tis amou (Gr), herzseeigel (G)

<u>Description/Biology:</u> The heart urchin is hard to find as it is usually buried in sand; however, its empty test can be found on the sea floor. This urchin is bilaterally symmetrical and has numerous soft thin spines giving it a brushy appearance. The spines are used for locomotion and keep sand away from the body of the heart urchin. It is active both during the day and night. It feeds selectively on organic particles in the sediment that are collected with modified podia on the oral surface. This heart urchin reaches a length of 10 cm and lives at depths between 3 m and 50 m.

<u>Classification:</u> Order Spatangoida, Family Schizasteridae

<u>Comments:</u> Heart urchins are probably not edible. Their test is more fragile than that of most other urchins.

Class Holothuroidea (sea cucumbers)

Sea cucumbers usually have a long worm-like body elongated along the oral-aboral axis. On one end of the body is the mouth surrounded by a ring of contractible tentacles. These mucous-covered tentacles assist in passing sediment and organic material to the mouth. Some holothurians feed from the substrate (deposit feeders). Others hold their tentacles in the water column to trap plankton (suspension feeders). On the opposite side from the mouth is the anus. In certain sea cucumber species, the anus provides protection for pearl fishes (Carapacidae). Some sea cucumbers have tube feet; however, all species lack arms and pedicellaria. The body surface of most sea cucumbers has a leathery texture. The body has the ability to change from being very flexible to being very rigid. When threatened or stressed, some holothurians expel a toxic or sticky substance that is used as a defence mechanism. Many holothurians have the ability to discharge part of their gut and associated organs. This phenomenon, known as evisceration, is followed by regeneration of the lost parts. The sexes are separate and in most species fertilization and development of larvae is external. Some species are brooders, incubating their eggs either internally or externally. Most sea cucumbers are benthic and can be found on a variety of substrates. Certain deep sea species are pelagic.

Some species are eaten in Asia after special preparation. The viscera and gonads are considered delicacies. Certain species are toxic. The toxins can usually be removed by boiling. Although sea cucumbers are not considered a food item in Greece, they are often used as bait for long line fishing.

Holothuria foskali

Common names: cotton spinner (En), angouri tis thalasas (Gr), forskals seegurke (G)

Description/Biology: When disturbed, the cotton spinner discharges sticky white or pink threads, known as cuverian tubules, from its anus. It can be various shades of brown with whitish rings. It can be found in sandy and rocky habitats from 1 m to 100 m deep. It is a deposit feeder, and is more active at night. The cotton spinner can reach a length of 30 cm.

Classification: Order Aspidochirotida, Family Holothuriidae

Comments: The sticky white discharge from the cotton spinner can stick to a diver or to his equipment for a fairly long time.

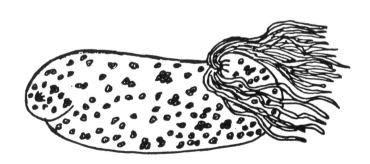

A sea cucumber releasing tubules of Cuvier

Holothuria tubulosa

<u>Common names:</u> brown sea cucumber (En), angouri tis thalasas (Gr), rohrenholothurie (G)

<u>Description/Biology:</u> The brown sea cucumber is very common in the Mediterranean. A deposit feeder, found over sandy and soft substrates from shallow waters down to around 50 m depth. Unlike the cotton spinner, it does not secrete cuverian tubules. The brown sea cucumber grows to 30 cm in length. It is active both during the day and the night.

<u>Classification:</u> Order Aspidochirotida, Family Holothuriidae

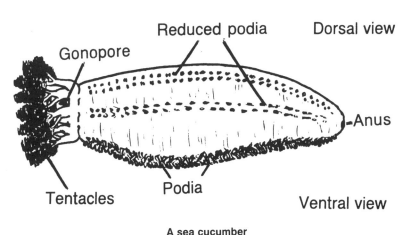

A sea cucumber

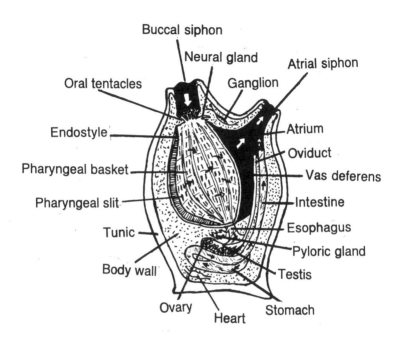

Cross section of a tunicate.
Arrows indicate path of water and mucous net.

Phylum Chordata (chordates)

Even though the majority (approx. 48,000) of species is this phylum belong to the subphylum Vertebrata (the vertebrates), approximately 1,400 species of chordates are invertebrates. All of these are marine species belonging to the subphyla Urochordata and Cephalochordata. Chordates at some point of their life have gill slits, a notochord, a nerve cord and a postanal tail.

Subphylum Urochordata (tunicates, sea squirts)

Ninety percent of the invertebrate chordates belong to this subphylum. The notochord and nerve chord are present in the larval stage but absent in the adult stage. Most are sessile filter feeders and there are no parasitic species.

Class Ascidiacea (sea squirts)

There are approximately 1,200 species (90% of the urochordates) in this class. They mainly inhabit shallow marine environments. Most are sessile and the majority of these are attached to hard substrates such as rocks, pilings and boat bottoms. The body is bag like and is protected by a secreted cellulose like substance called tunic. Ascidians are filter feeders. Water is drawn into the body through an anterior buccal siphon and is pumped out through a smaller atrial siphon. Water current is generated by the beating of many cilia. Food is strained and passed to the stomach and it is excreted close to the atrial siphon. Some ascidian species are solitary and others are colonial. Colonial species usually share an atrial siphon and reproduce asexually by budding. Solitary species reproduce sexually and most are hermaphrodites. Cross fertilization is much more common than self fertilization. In solitary species the eggs are planktonic, whereas in colonial species the eggs are usually brood in the oviduct or atrium. When disturbed, many ascidians can contract and close the openings of their siphons.

Halocynthia papillosa

Common names: red sea squirt (En), askos (Gr), rote seescheide (G)

Description/Biology: The red sea squirt is solitary and usually grows to be about 10 cm in length. It attaches to hard substrates at depths between 3 m and 50 m. This ascidian is a diurnal filter feeder and reproduces sexually. The larval stage is planktonic.

Classification: Order Stolidobranchia, Family Styelinae

Comments: Even though this ascidian is not known to be edible, some other species are considered a delicacy.

Laws and Regulations for Recreational Fishing

General Restrictions

It is illegal to fish at night with the aid of light sources. The only exception is fishing with a hand held spear using a light source of less than 500 candle power. Fishing in marine reserves and aquaculture areas is prohibited. Hand line, and rod and reel are the only types of fishing allowed in ports. It is illegal to fish for corals and sponges. The minimum weight requirement for octopus and lobster is 500 g. Selling catches is also prohibited. Sportfishermen and sportfishing vessels require a license for recreational fishing.

There are a number of open and closed seasons, especially for certain marine invertebrates such as octopus and lobster. These seasons may vary from area to area and therefore individuals should consult the Fisheries Department or local port authorities to obtain accurate and up to date information.

Spearfishing

A minimum age of 16 is necessary to spearfish in Greece. While engaged in spearfishing, the fisherman's location should be indicated by a yellow buoy with a yellow flag with a red diagonal line. If diving from a boat the flag may be located on the boat. The flag must be visible from at least 300 m and the spearfisherman should operate inside an area of 50 m radius from the flag.

It is illegal to spearfish after sunset and before sunrise, during the month of May, at the entrance of ports, in areas of boat traffic and within 200 m from bathing areas and marked nets. The use of breathing apparatus such as SCUBA while spearfishing is illegal. The minimum size requirement for fish is 150 g. In addition it is prohibited to remove fish and other creatures from nets, long lines, etc. The use of chemicals to aid fishing is strongly prohibited. Spearguns should only be loaded in the water.

Nets and Long Lines

The use of long lines and nets is prohibited for the month of May. A long line with a maximum of 150 hooks per person or 300 hooks per fishing vessel is allowed. The maximum allowable length of nets is 100 m, the eye diameter should be greater than 24 mm. All nets and long lines should be marked with buoys that are visible from at least 300 m. The total maximum weight of catch should not exceed 10 Kg; however, it is legal to keep an individual fish weighing more than 10 Kg.

Cast Nets and Fish Traps

One cast net of maximum diameter of 10 m is allowed per person. The eye diameter should be larger than 30 mm. It is illegal to fish with cast nets in estuaries or within 500 m from the mouth of rivers and within 200 m from aquaculture areas. Two fish traps are allowed per fisherman with an eye diameter larger than 40 mm. Lobster traps are illegal.

Overfishing, Pollution and Conservation

Invertebrates such as lobsters and octopus attain high market prices. For this reason they are actively sought after by fishermen and subject to intense fishing pressure. There are several different ways that marine invertebrates are fished. Some of these fishing techniques are more destructive than others. Dredging and bottom trawling can have several negative effects on the benthic community. Richly structured benthic habitats are transformed to less structured environments that are incapable of sustaining a majority of the original fauna. Disturbed bottoms have an increased amount of suspended particulate matter which can clog the feeding apparatus of many invertebrates and reduce light penetration. Many fishes and invertebrates are not able to colonize, feed or avoid predators in these disturbed environments. Other fishing methods, such as lobster potting which is currently illegal in Greece, are too efficient and can deplete an area of most of its targeted resources.

Even though fishing with explosives such as dynamite is illegal, it is frequently used in remote areas to kill fishes. The effects are devastating not only to all the fishes but also to the marine invertebrates and flora in the area. Explosives can transform a luscious and rich productive area to an underwater desert void of any visible life for many years.

The illegal use of chemicals to capture invertebrates and fishes poisons the surrounding area and kills a wide diversity of organisms. The affected area may take a very long time to recover. Unfortunately many sport fishermen use copper sulphate solution to remove octopuses from their hiding places or chlorine to capture fishes.

Many invertebrates are active at night and thus are very vulnerable to spearfishing at this time. In addition, a diver using SCUBA can spend a substantial time underwater "collecting" invertebrates. This can have detrimental effects on targeted species in an area. For this reason, the use of spearfishing at night as well as the use of SCUBA are illegal.

As the tourism industry in Greece has grown, coastal areas have become more developed. This has resulted in an increase in coastal organic pollution and trash, specifically during the summer months due to the influx of a large number of tourists. Organic pollution can promote algae blooms which at times can be toxic to vertebrates and invertebrates. In addition water turbidity can increase, reducing the maximum depth at which photosynthesis can occur. Non biodegradable trash such as plastics can smother benthic communities and drastically reduce light and food availability and thus starve them to death. Invertebrates that can move may be able to avoid death by relocating.

Even though the Mediterranean generally is considered to be polluted, Greek waters away from industrial areas and major cities are usually very clean. Heavily polluted areas in Greece are those around Athens and Pireaus (Saronikos Gulf), Thessaloniki (Thermaikos Gulf) and Patra (Patraikos Gulf). These areas receive both inorganic and organic pollutants from industrial belts that surround the cities as well as river outputs (Thermaikos Gulf) that carry fertilizers/pesticides from agricultural areas.

Since ancient times, the sea has played a vital role in daily life in Greece. Greece relied on the sea as a provider of food, a means of communication and as a source of cultural inspiration. Now more than ever before, Greece relies on the sea for its well being as a part of the highly specialized European Community. Because of its flourishing tourism and its rapidly developing aquaculture, Greece needs to maintain clean and healthy seas.

Strict enforcement of the existing laws and regulations is crucial in order to sustain a rich marine environment. Unfortunately, due to limited resources and manpower, it is extremely difficult to enforce the existing laws and regulations along the vast coastline of Greece. Therefore, we as individuals must make a conscious effort to protect the Greek marine environment.

Aquaculture

Over the last decades the aquaculture of invertebrates, mainly molluscs, has increased dramatically. Molluscan aquaculture is generally low cost, efficient and environmentally friendly. Most cultured molluscs are herbivorous and utilize plankton that is directly unavailable to humans. The majority of cultured molluscs are bivalves from temperate estuarine areas. Temperate species are more hardy than their tropical counterparts, making them better able to withstand salinity, temperature and desiccation stresses.

The first documented effort of mollusc culture dates back to the time of ancient Rome, when a nobleman noticed that by providing substrate he could increase the production of oysters. This simple, primitive method of aquaculture is still used in some areas today. Japan is currently one of the world leaders in aquaculture, both in species diversity and quantity.

Today in Greece there are over 100 aquaculture farms that produce invertebrates. The majority of them produce mussels, while others produce oysters. Scallop and crustacean culture are not as developed yet but show high potential.

In contrast to molluscan aquaculture, crustacean culture uses tropical species and often has detrimental ecological effects. For example, some shrimp culture techniques in Asia result in the destruction of mangroves which are important nursery grounds for many marine fishes and invertebrates. Most crustacean culture is not very intensive. It is usually concentrated on brackish shrimp species and crayfish. Asia is the dominant producer of cultured shrimp.

Shellfish aquaculture in Greece is still secondary to fish aquaculture. Recently an old form of aquaculture is becoming more popular. By incorporating different types of aquaculture and even agriculture into what is known as polyculture or integrated aquaculture, a more sustainable means of food production is achieved. For example, fishes such as tilapia and carp are often cultured in rice paddies, and mussels are grown just outside of fish pens. Both of these techniques are an efficient means of utilizing excess nutrients for further food production.

Invertebrates in medical research and as a source of biochemical compounds

Secondary compounds from organisms have been used by shaman, witch doctors and the biochemical industry to treat a number of illnesses. Until recently terrestrial plants were the usual source of biochemical compounds. In the last decades as exploration of the sea advanced, biochemists started looking to marine organisms for new substances. Even though this field is in its primary stages, the prospects appear promising. Private companies from countries such as the USA, France, Japan and Spain have shown interest in marine biochemical compounds and some research cruises have been exclusively devoted to examining marine organisms for potential new compounds.

Sponges - Even though sponges are usually considered to be mostly used as cleaning aids, many compounds have been isolated from a diversity of sponges. These compounds are used primarily by pharmaceutical industries. Some of the properties of these compounds are antileukemic, antiviral, anthelminthic, antiinflammatory and analgesic.

Cnidarians - Compounds with cardiotonics have been isolated from several cnidarians. In addition, some species possess powerful antifouling compounds; however, to our knowledge, these have not yet been used by the industry.

Molluscs - Nudibranchs have been found to contain compounds with antifungal and antileukemic properties. Some prosobranchs posses secondary compounds with antiviral, antimicrobial and antitumor activities.

Annelids - Some worms posses antifungal, antileukemic and insecticide-like compounds.

Crustaceans and chelicerates - Horseshoe crabs have been used for cardiovascular and circulatory research.

Ascidians - Some sea squirts possess substances that enhance antibody production. These compounds are used as anticancer agents. Members of the sea squirt family Polychitoridae contain substances that have antiviral properties.

Echinoderms - The pigments in the spines of sea urchins have been found to have antibacterial properties and sea urchin eggs have been used to test bioactive products. Antipredator compounds of some echinoderms have antimicrobial, antiviral and cytotoxic activities.

Glossary

Abyssal depths- depths between 3,000 m and 5,000 m

Ambulacrum- groove, ridge or band of tube feet, radial canal, and the associated body wall of echinoderms

Anterior- situated towards the front

Aperture- a hole or opening

Asexual reproduction- reproduction without the formation of gametes

Behaviour- the response and action of an organism to external and internal stimulation

Benthic- relating to the bottom

Bilateral symmetry- one plane of symmetry divides the organism into two mirror image halves

Bilobed- consisting of two lobes

Bioaccumulate- the accumulation of a substance in an organism at a greater concentration than that at which it occurs in the inorganic environment

Bioindicators- organisms that are used to assess the condition of an environment and/or environmental changes over time

Bioluminescense- light that is generated biochemically and is emitted by organisms

Byssal threads- tough protein threads, produced by a gland in a the bivalve foot, that are usually used for attachment

Calcareous- made up of calcium carbonate

Carapace- the exoskeletal plate that covers at least part of the anterior dorsal side of many arthropods

Carnivore- an organism that eats other animals

Cartilaginous- made up of cartilage

Cephalothorax- the combined head and thorax

Chaetae- bristles made up of mostly chitin; characteristic of chaetopod Annelida (Polychaeta, Oligochaeta) in which they occur in groups projecting from the skin.

Chelate- refers to claw-like or pincer-like appendages, usually those of arthropods

Cheliped- a clawed or chelate thoracic appendage of decapod crustaceans

Chromatophore- pigment cell in the body wall that can contract and expand to expose and conceal its pigment

Cilium (pl. Cilia)- a motile outgrowth of the cell surface that is typically short and has an oarlike stroke

Cirrus (pl. Cirri)- tentacle-like and curved appendages found in different animal groups

Coelom- body cavity that is lined by a mesodermally derived epithelium

Compound eyes- image forming eyes of many arthropods composed of multiple lenses and photo receptors called ommatidia

Congeneric- belonging to the same genus

Conspecific- belonging to the same species

Corallite- skeletal parts deposited by a single polyp

Cosmopolitan- organisms with worldwide distribution

Cuverian tubules- white, pink or red tubules that shoot out of the anus in some species of sea cucumbers as a defence mechanism

Deposit feeder- an organism that obtains its nutrition by consuming some fraction of soft sediment

Detritus- fragmented decomposing organic matter

Dioecious- having separate sexes

Diurnal- active both during the day and the night

Dorsal- on or near the upper part of the organism

Ectoparasite- a parasite that lives on the outer surface of the host

Encrusting- the covering of a surface area with a foreign material or organism

Epidermis- outer epithelial layer of the body (skin)

Epipelagic- relating to the upper 200 m of the ocean

Euryhaline- organism capable of living in waters of a wide salinity range

Eutrophic- water bodies or habitats having high concentrations of nutrients

Evisceration- when the anterior or posterior end of an organism ruptures and parts of the gut and associated organs are expelled

Family- group of organisms with similar morphological characteristics

Fauna- the entire animal life of a given region or habitat

Filter feeder- an organism that feeds on particles such as plankton and detritus by filtering the water

Flagellum (pl. Flagella)- a long outgrowth from many protozoan and metazoan cells that has a complex and whip-like motion

Gamete- a mature haploid reproductive cell that fuses with another gamete of the opposite sex to form a zygote

Gastrodermis- cellular epithelial lining of the gastrovascular cavity of cnidarians and ctenophores and the midgut lining of bilaterally symmetrical animals

Gastrovascular cavity- internal extracellular cavity of cnidarian and ctenophores lined by gastrodermis

Grazer- an organism that consumes organisms far smaller than itself

Gregarious- tending to aggregate actively into groups or clusters

Habitat- the place where a species usually lives

Herbivore- an animal that feeds on plants

Hermaphrodite- an animal which has both male and female sex organs

Interambulacral- areas between ambulacral plates that are devoid of podia, found in echinoderms

Intertidal- the area on the shore that is between the high and low tide levels

Invertebrate- organisms that do not have a back bone in any part of their life

Keel- a raised ridge, usually sharp

Larvae- the early life stages of invertebrates and fishes

Littoral- relating to near-shore waters

Lophophore- a food trapping circular ridge that encircles the mouth and bears many hollow ciliated tentacles

Luminescence- see bioluminescence

Mantle- body wall that may secrete a shell, found in molluscs, brachiopods and ascidians

Mesenteries- longitudinal sheets of tissue that divide the body cavity of bilaterally symmetrical animals

Mesopelagic- relating to the portion of ocean below the epipelagic zone to 1000 m

Metabolites- substances that take part in the processes of metabolism

Microhabitat- a small specialized habitat

Migration- movements of animals from one location to another, usually on a regular basis

Molting- shedding of the old cuticle as a new cuticle is being secreted

Nacreous- pertaining to the innermost, lustrous, shell layer of molluscs

Nematocyst- a capsule containing a thread-like stinger used for anchoring, defense, or capturing prey

Neritic- pertaining to the waters over the continental shelves

Niche- the special habitat that supplies all life-controlling factors (physical, chemical and biological) to an organism

Nocturnal- relating to night activity

Nodular- having hard or swollen junctions

Omnivorous- organisms that feed on both plants and animals

Operculum- lid; covering flap

Ouzo- Greek aniseed flavoured alcoholic beverage

Oviparous- producing eggs that develop and hatch outside the body

Parapodium (pl. parapodia)- fleshy, segmental appendage of polychaete worms

Parasites- organisms that live on or in the host and benefit at the expense of the host

Pedicellaria-a small specialized jaw-like appendage of asteroids and echinoids that is used for protection and feeding

Pentamerous- a type of symmetry consisting of five parts

Pelagic- relating to the open sea, outside littoral waters

Pharynx- the anterior part of the gut

Photophobic- intolerant of, or avoiding, bright light conditions

Pinnate tentacles- tentacles that have side branches, like a feather

Podium (pl. podia)- tube foot of an echinoderm

Polymorphic- the co-occurrence of several different forms of the same species

Polyp- an individual of a solitary coelenterate or one member of a coelenterate colony

Posterior- situated towards the rear

Plankton- free-floating organisms from the sea, usually microscopic

Predator- a carnivorous animal that feeds by killing other animals

Proboscis- tubular protrusion of the head, snout, or anterior part of the gut; usually used in feeding and often extensible

Protandry- a condition in hermaphrodites in which the male gonads are sexually mature before the female gonads

Protogyny- a condition in hermaphrodites in which the female gonads are sexually mature before the male gonads

Protractile- capable of being thrust out

Radial symmetry- the arrangement of similar parts around a central axis so that any plane through this axis will divide the organism into similar halves

Radioles- tentacles on the head of some polychaetes

Radula- horny toothed organ of molluscs; used to rasp food and carry it to the mouth

Retsina- a Greek white wine that has a slight resin flavour

Salinity- a measure of the total amount of dissolved salts in sea water

Scavenger- an animal that feeds on the remains of other animals or plants

School- large organised aggregations of marine animals (usually, fish of the same size and species)

Sedentary- an organism with a small home range that does not migrate

Septa- double-walled tissue partitions in the cross-sectional plane of bilateral organisms or a radial plane of cnidarians

Serration- a formation resembling the toothed edge of a saw

Sessile- attached to the substratum and thus immobile

Sexual dimorphism- the structural or colour differences between sexes of a species

Sexual reproduction- reproduction that requires the fusion of male and female gametes to form a zygote

Solitary- alone; not part of a colony or school

Spine- a bony structure, usually stiff and sharp-tipped

Species- a group of closely related individuals which can and normally breed to produce fertile offspring

Spicule- a small needle-like or rod-like skeletal piece

Strobilation- a type of asexual reproduction found in scyphozoans

Stylet- a dagger-like structure associated with various systems of different animal groups

Substrate- the base on which an organism lives

Sublittoral- the marine zone extending from the lower margin of the intertidal (littoral) to the outer edge of the continental shelf at a depth of about 200 m; sometimes used for the zone between low tide and the greatest depth to which photosynthetic plants can grow

Subumbrella- lower oral surface of a medusa

Suspension feeders- organisms that feed by trapping particles suspended in the water column

Symbiotic- an intimate physical relationship between two species in which at least one of them is dependent upon the other (to various degrees); parasitism, commensalism, and mutualism are the three types of symbiotic relationships

Synchronous hermaphrodite- an animal that has mature male and female sex organs at the same time

Siphon- an accessory gut of echiurans and some echinoids; a tubular fold of the molluscan mantle used to direct water to and from the mantle cavity; inhalant and exhalant apertures of urochordates

Tentacle- evagiantion of the body wall surrounding the mouth which aids in the capture and ingestion of food

Test- shell, hard covering, valve or theca

Trochophore- type of larva found in molluscs, annelids and other groups in which the larval body is ringed by a girdle of cilia

Truncate- square cut

Trunk- body

Tube feet- hydraulically controlled feet, part of the water vascular system of echinoderms

Ventral- on or near the lower part of the organism

Zoid- colonial individual that resembles but is not a seperate organism

Zooplankton- animal members of plankton

Zooxanthelae- symbiotic cyanonacteria usually found in association with corals

Index of Common Names

Index of Scientific Names

About the Authors

Ioannis Eystratios Batjakas was born on July 18, 1964 in Mytilene, Lesvos, Greece. Ioannis graduated from T.E.I., Messolongi, Greece, in 1986, earning a Diploma in Fisheries and Aquaculture. He then went to United States where he graduated from Boston University, Boston, MA, U.S.A., in 1989, earning a Bachelors Degree (cum laude) in Marine Biology. Furthering his studies at the University of Massachusetts Dartmouth, MA, U.S.A., in 1994 he earned a Masters of Science Degree in Marine Biology, specializing in ichthyology and animal behaviour. His Master's thesis was a comparative feeding behavioural study of two species of cichlids common in Lake Victoria, Africa. He completed his doctoral studies in 1999 at the Boston University Marine Program, Woods Hole, MA, U.S.A. During the last 4 years he has also worked as a scientific consultant for several organizations on a variety of projects. Now he is conducting research at the Fisheries Research Institute, Kavala, Greece.

Alistair Evelpidis Economakis was born on May 21, 1971 in Athens, Greece. He graduated in 1994 from Boston University, Boston, MA, U.S.A. with a Bachelors Degree in Marine Biology and Environmental Science. He completed his Masters degree in 1995, specializing in shark behaviour and ecology. In 2000 he completed his doctoral studies in marine biology at the Boston University Marine Program, Woods Hole, MA, U.S.A. Alistair used cichlid fishes as a model to investigate the impacts of behaviour and individual variation on resource utilization patterns. During his studies, Alistair has worked for the Woods Hole Oceanographic Institute on several research projects at Johnston Island in the Mid Pacific. In addition, he has worked as a research assistant and wildlife photographer at Lake Victoria, Kenya.